AT VARIANCE

*The Church's Argument
against Homosexual Conduct*

AT VARIANCE

The Church's Argument against Homosexual Conduct

The Revd Dr Kevin F. Scott
Rector of St Philip & St James' Church, Edinburgh

Published with support from
The Scottish Order of Christian Unity

DUNEDIN ACADEMIC PRESS
EDINBURGH

Published by
Dunedin Academic Press Ltd
Hudson House
8 Albany Street
Edinburgh EH1 3QB
Scotland

ISBN 1 903765 37 4

© 2004 Kevin F. Scott

The right of Kevin F. Scott to be identified as the author of this work has been asserted by him in accordance with sections 77 & 78 of the Copyright, Designs and Patents Act 1998.

All rights reserved.

No part of this publication may be reproduced or transmitted in any form or by any means or stored in any retrieval system of any nature without prior written permission, except for fair dealing under the Copyright, Designs and Patents Act 1998 or in accordance with the terms of a licence issued by the Copyright Licensing Society in respect of photocopying or reprographic reproduction. Full acknowledgement as to author, publisher and source must be given. Application for permission for any other use of copyright material should be made in writing to the publisher.

British Library Cataloguing in Publication Data
A catalogue record for this book is available from the British Library.

Printed in Great Britain by Cromwell Press.

Contents

Foreword by the Revd David W. Torrance		vii
1	*Introduction*	1
2	*Old Testament Judaism*	10
3	*Jesus and the Gospels*	22
4	*The Letters of St Paul*	36
5	*Some Representative Writings from the Patristic Period*	48
6	*The Church's Plight*	56
7	*Casualties and Consequences*	64
Appendix 1: St John Chrysostom's Homily on Romans 1:26–27		75
Biblical Index		84
Subject Index		86

Foreword

This is an excellent and much needed little book and Dr Kevin Scott has done great service to the Church and society in writing it. His approach is thought provoking. He has written with clarity, sensitivity and in an irenic way. Probably no other issue today so troubles and even threatens the unity of Churches in the Western world as that of homosexuality. It is also much debated in society at large. It has been and is the subject of legislation by governments.

The book has real relevance for both individuals and communities, especially in our day when the cry is for openness. It is a most valuable contribution in this field, touching areas in both Church and state that look for answers and conclusions.

As members of the Scottish Order of Christian Unity we warmly commend this book to all who seek greater understanding of this subject and of the moral and spiritual implications for Church and society of homosexual practice.

The Revd David W. Torrance
Chairman, Scottish Order of Christian Unity

The Scottish Order of Christian Unity is an ecumenical Order. It seeks *inter alia* 'to encourage all Christian people to be faithful to the teaching of Christ' and to promote 'essential Christian ethics for the common good' in both Church and society.

1
Introduction

This book sets out briefly the principal arguments of Classical Christian Orthodoxy[1] against homosexual conduct. But it is as well to lay out the grounds and define the terms before we start. Many of the engagements on this subject within the modern Church are at least partly vitiated by a lack of argument. One of the reasons why it took several hundred years of thought and discussion in the Early Church to produce the Nicene Creed, which is the universally acknowledged statement of the essentials of Christian belief, is that doctrine had to be argued from the available evidence, the case had to made for understanding each part of the Christian profession. While Scripture commands respect as the primary authority governing Christian belief and conduct, it is insufficient merely to quote it as a kind of *fiat*. What is written in Scripture may have great authority in establishing what is true and what is not, what is permissible and what is disallowed, but it is not opaque to rational enquiry. Christianity is reasonable: it is possible to comprehend the statements and declarations of Scripture in

---

1 'Classical Christian Orthodoxy' is used to mean the doctrinal and moral consensus founded on Scripture and expressed in the Great Councils of the ancient Church and in the writings of the Church Fathers. The consensus that these writings and the deliberations of the Councils represent, has, in turn, been adopted and applied by the worldwide Church up to the present day with a high degree of unanimity. The definition of orthodoxy as that which is held and believed *semper, ubique ab omnibus*, always, everywhere and by everyone, is an ideal not absolutely sustained in every case, but has held with remarkable breadth of agreement until modern times.

terms of the nature of God as he has revealed himself, and in terms of the redemptive work by which he saves and restores humanity to himself. This means that Christianity can be properly *understood*. It is not a kind of spiritual pill which needs to be swallowed, but a way of learning by which Christians can make sense of their lives and live in accordance with the truth.

Correspondingly, its precepts cannot be abstracted from the whole and applied without reference to their source and meaning. One of the great difficulties of our modern world (particularly in the West) is that the remnant of Christian morality is being applied to the Church, while a secular world that knows nothing of the Christian foundations, which are essential to its comprehension and acceptance, looks on with an ever increasing sense of remoteness. For representatives of the Christian Church to reel out Christian prohibitions to an increasingly bewildered secular audience fails to advance the Christian argument and tends to deepen the chasm between the two. Equally, inside the Church, many Christians have a very poor understanding of their own faith and may be just as ignorant of the arguments of Christianity as are those outside.

What is really necessary is a recovery of the will to argue the Christian case within the Church, so that Church members hold their faith with confidence in what stands behind it and with a critical recognition of the arguments against it. Similarly, the case for Christianity and for its moral consequences should be set out to the world so at least people may understand the basis for the argument, even if they cannot always accept it.

It is the intention of this book to outline the arguments against homosexual conduct not simply by resorting to the textual *fiat* of Scripture nor to the fact of the traditional opposition of the Church to such activity, but by arguing from the first principles implicit in Christianity and in the Judaic tradition which stands behind it. It will not simply be a restatement of Biblical law or merely a demonstration

Introduction

that homosexual practice has never been admissible in the past. Rather, it will seek to show that both Jewish law and Christian moral obligation rest on the foundations of Judaeo-Christian faith, upon the revelation of God and of His saving acts. The book is thus intended to be *reasonable* in the proper sense of the term. It is not apologetic in that it makes no attempt to produce some compromise or 'third way', in which the teaching of the Church can somehow be rationalised with a permissiveness towards homosexual practice. Properly understood it is *polemical*[2] in that it takes a definite stance against homosexual conduct and hopes and expects to win the argument or, at the very least, aims to present to the reader points which will have to be answered before homosexual practice could be thought of as permissible. The book may, therefore, raise ire, although that is not at all the intention. It is intended to shed light and clarify the terms of the argument. It may be thought to be 'homophobic'. The charge of homophobia tends to have the effect of blocking argument. Once the accusation is raised the terms of discussion tend to change. The accused are tempted to defend themselves against the charge at all costs and the force of the original argument may be lost. May it suffice to say that the intentions of this book are to present the argument with clear rationality, with an absence of rancour, with a desire to edify the Church and to assist Christians in their pilgrimage, and without malice towards any opponent.

It is at this point that we must identify clearly and exactly that which is being opposed. Throughout this book, an attempt will be made to use words with as much precision as possible. 'Homosexual' is one such word, and some space is given here to its meaning. In a

---

2 The word is often used in modern parlance to suggest a cheap attack at the expense of truth, and thus not worthy of proper consideration, but properly 'polemic' or 'polemical' do not carry such connotation. The Church Fathers often wrote polemically against an idea or position inimicable to Christianity, but they argued their cases with considerable thoroughness.

modern dictionary, the term is usually defined thus: *of, relating to, or having a sexual orientation to persons of the same sex*. Its use as a noun is sometimes regarded as pejorative, but not so much as an adjective.[3] The above definition will not be used in this book, however, because it is not *orientation* which is being challenged but *sexual practice*. A detailed discussion of the terms used in the New Testament will be postponed to a later chapter, but words sometimes translated as 'homosexual' refer to individuals who perform homosexual genital acts. This is the definition adopted in this book. As far as this book is concerned, a homosexual is not someone who has this or that inclination or orientation, but someone who performs sexual acts with another individual of the same sex. There is an important distinction being made here, which will be discussed further later on but also needs to be emphasised at this stage. The arguments presented here are not an attack upon cultural or social matters or upon artistic preferences. Nor are they a simplistic treatment of the complex issues of femininity or masculinity. To a very great extent, the argument is limited here to what sexual acts it is permissible or not permissible to perform according to Judaeo-Christian ethics and why. For clarity, in this book the word 'homosexual' will only be used adjectivally. The terms 'homosexual conduct', or 'homosexual practice' or 'homosexual activity' will all be used exclusively to mean 'homosexual genital acts'.

By restricting the definition in this way, we raise the question of whether sexual conduct is sufficiently important for the Church to make it such a serious issue. What does it matter if such things are confined to people's private lives? A fuller answer to this question will be given in later chapters, but here it is worth noting that from the

---

3   The use of the word 'homosexual' as a noun is now considered politically incorrect because of the emphasis the term places upon sexuality rather than on cultural and social matters. 'Gay and Lesbian' are apparently preferred for this reason. The word 'homosexual', both as a noun and as an adjective, will be used in this book precisely because sexual conduct alone is being challenged.

Introduction

beginning both Christianity and Judaism understood their theology to penetrate to everyday life at ground level. The highest notions about God had a governing impact on daily living, on eating, drinking and marrying, on relationships with family members, neighbours, and the human race in general. For Jews and Christians alike, faith is not the adherence to a set of credal formulae or the observance of arcane religious practices; rather faith is a *way*, both a path of life, and a manner of being in the world. The key thought is that this way of life must make sense in terms of the divine character and the nature of creation and must be consistent with God's will and purpose, and pleasing to him. In Chapter 2 we will examine this relationship in the life of Ancient Israel.

The creative and redemptive aspects of God's dealings with Israel are further developed in the New Testament in the life and ministry, death and resurrection of Jesus. These things are considered in Chapter 3, where the role of Jesus Christ, in effecting in himself the return of Israel from Exile, forms the foundation of moral instruction in the four Gospels. Here the author of the present book is indebted to N. T. Wright's *Jesus and the Victory of God* for his lucid interpretation of Jesus' ministry in terms of the Old Testament and Later Judaism. Importantly, this allows a rational continuity of thought between the moral implications of the Old Covenant and the striking ethical demands of the New.[4] But recognition is also given to the fact that Jesus did not merely prescribe the moral necessities of the Kingdom but exercised them. He comes among his people with the infinite demand of God coupled with His absolute acceptance and with His limitless power to address their physical moral and spiritual plight. We will examine two instances, relevant to the subject of this book, in

---

4 That continuity is, of course, not dependent on the details of Wright's treatment, particularly his argument that Jesus effects Israel's Return from Exile, but the general principle that Christ was restoring Israel is the basis for theological and moral continuity between the Testaments.

which these things are brought to bear upon human affliction in a decisive way.

In Chapter 4 the argument moves to the writings of St Paul, first with a treatment of the theology underlying his understanding of homosexual conduct from Romans 1. Following Hays, we consider the charge that homosexual conduct is a kind of anti-sacrament of man's primordial refusal of God. We then demonstrate how Paul argues that all sins are manifestations of the disordered life of fallen man, and that the Christian Church must be understood as the community of Jesus Christ, forgiven and restored, and which is now disqualified from returning to the things from which she has been delivered. This will mean that practices considered acceptable outside may not be at all acceptable within the Church, as the nature of the Gospel might lead us to expect. Indeed, any reading of the life of Jesus in the Gospels might cause us to anticipate a dramatic contrast between Christian living and the secular world. This will include a striking contrast between the world's attitude to homosexual conduct and that of the Church. This contrast is reflected in St Paul's ethical instructions to the Churches, and shows the intrinsically adversative relationship that he expected between the Christian community and the world, in terms of world-view and morality. Despite changes in social and political conditions from age to age, including the abolition of slavery, the equalisation of the sexes, etc., that differentiation remains because it is founded on the transforming impact Jesus Christ had on those who received his ministry.

Chapter 5 takes us further into that contrast through the writings of some of the Church Fathers. Here we find a robust atmosphere very different from the cautious restrained air of our own day. The Patristic authors do not merely think that homosexual activity is morally incoherent with Christianity, they think it is madness itself, an insanity of diabolical proportions, a sin worse than murder! But like everything else, they do not simply declaim this, they argue it from Judaeo-

Christian first principles. It might be thought that their stance is too extreme, excessively rigorous, but it cannot be argued to be irrational. In fact it is the very opposite: they occupy the position they do because they allow the Judaeo-Christian argument to lead them where it wills. We live in an age that looks ahead in any argument to see if it likes the conclusion; if it does not, it quickly closes down the argument. The Church Fathers did not do this. But because of the route they take, their final position, albeit extreme by modern standards, can actually be well argued and staunchly defended, which is again another surprise for us: we are inclined to think that any extreme argument must be weak.

Chapter 6 begins to investigate a new thesis. We will have already seen in Chapter 4 that St Paul regards homosexual conduct as an anti-sacrament: an outward, visible sign of an inward rebellion against God and a descent into mental and moral chaos. According to Paul, this took place primordially, a declension of the human race outside the reach of history, but which extends its destructive power throughout the generations. But within history, in fact, relatively recently, there has been a parallel declension of the mind of the Church which has become darkened and discredited in a particular way. Over the last hundred years at least, the Church has admitted Feuerbach's reductionism as a legitimate mode of reasoning. This reductionism tells us that God is not the external, independent holy Creator who acts independently to save his creatures. According to Feuerbach, God is nothing other than a psychological projection of man, the entirely speculative assertion of an Absolute, which has no independent or real existence except in the imagination and 'faith' of the religious person. This is, of course, an atheistic position, but it has been embraced by the Church to such an extent that many so-called theological works have been based on it, and many clergy trained to understand faith that way. It forms the basis, at least in part, of the 'liberal' position in the Church. It is contended here that this way of thinking is dismissive of

God, does not give God proper credit and is, thus, exemplary of a discredited mentality. According to St Paul, homosexual conduct is the outward evidence and consequence of this type of discredited mentality. Accordingly, it is not now surprising that the Church is ready to embrace homosexual activity as normative. If we had been astute enough we could have predicted it a hundred years ago. Apart from helping us to understand why we are having this debate in the Church just now, this analysis does enable us to see the direction we must take to escape the present predicament. It will not be sufficient to engage in penitence, merely taking the form of a re-establishment of Judaeo-Christian sexual morality, but also, more importantly, we need to recover a genuine Christian way of thinking about theology.

Finally, in Chapter 7 we consider the casualties and consequences of the legitimisation of homosexual practice. We will have already considered the anti-theological, anti-redemptive and anti-sacramental features of this fatal step. We now consider its impact on the unity of the Church, the prized spiritual, moral and theological agreement which has been preserved with such great difficulty across the ages. We question how we could maintain our communion with the Church Fathers and all Christians in previous ages if we were to pronounce what they understood to be a mortal sin, henceforth a virtue. We ask about the doctrine of the Cross and of the forgiveness of sins. If this sin is a sin no longer, the question is raised of what the Cross achieved and whether forgiveness is now not a variable factor from age to age. These are anti-ecumenical and anti-penitential factors. Then there is the casualty of rationality. If we now authorise things which have no meaning or coherence, we enter the world of the anti-scientific and anti-educational. If these things have no basis for social order, they are also essentially antisocial, and because they dissolve meaning from action, their assumption will mean that the Church can no longer calculate the pastoral good of its members. The adoption of homo-sexual practice in the Church can only be understood as anti-pastoral.

Introduction

We have now given some shape to the argument set out in the subsequent pages. The coverage is by no means comprehensive, but the principal lines of reasoning are given whereby a sufficient account may be made of all the main references in Scriptures and the Church Fathers to homosexual conduct in the context of the theology of the Judaeo-Christian tradition, to which we now turn.

2
Old Testament Judaism

If we are to understand the teaching of Ancient Israel on the issue of homosexual practice we need to start a long way back. We cannot simply quote the texts of the Old Testament and hope to justify the support of Church members and satisfy the criticisms of those outside. In this chapter, we will not begin with the start of the Bible or with pre-history in the Creation narrative, but with Israel's formative discoveries of the character and mode of action of the God who revealed himself to her.

Setting aside the revelation of God to the Patriarchs, Israel as a nation came into being as a result of the astonishing liberation of the Hebrews from slavery in Egypt, which Israelites could only attribute to the saving action of a God who was new to them and whose love they recognised but could not explain. The uniqueness of the faith that arose as a result of their liberation has been well attested.[1] In their experience of deliverance from Egypt, their wilderness wanderings, their entry into and settlement in the Promised Land, they slowly comprehended something of the character of the God who had saved them. They learned first that his love for them was *redemptive*. It took the form of a rescue in which they were taken from a place of total disadvantage, poverty, misery and disarray, and brought to a place of great advantage, riches, happiness and good order.

---

1 For the uniqueness of Israel's faith compared with ancient near eastern religions see, for example, Roland de Vaux, *Ancient Israel*, vol. 2 (New York: McGraw Hill, 1961).

They also discovered that there was a great differentiation between God and them, and between God and everything else. It was not possible to extrapolate from observations about themselves or the world in general to gain any kind of secure understanding of God. Although they believed themselves to be united with God through his promise, they came quickly to know that he was utterly different from them and, importantly, different from any other ideas people might have about deity in general. The shorthand for this divine differentiation is *holiness*. Under this principle, they could not look around at other nations and their religions, put together their own version and call it theology and impose it, as it were, on God. The information was to run quite the other way. God had revealed and would continue to reveal himself to them. They would come to a knowledge of God, not through their own capacities but by divine revelation.

Notice that the redemptive character of his love and his holiness are both characteristics which imply the drawing of distinctions. They had been in Egypt, they were now in the Promised Land. They had been without God and without hope in the world, now they were God's people with a unique destiny. Once they had been nobodies, a law unto themselves or subject to the capricious will of Pharaoh, now they were the people of the holy God and subject to his ordering governance.[2]

It is not clear at what stage Israel began to think seriously about other theological questions, such as the issue of Creation and cosmological matters. It may have been during the Babylonian Exile where they were confronted by speculations concerning the origin of the world and attempts to explain the paradoxes of human existence, or it may have been much earlier. Either way, when they were asked who made the world, they would readily acknowledge that it was the God

---

2   In the first letter of Peter, the Christian Church is described in exactly these terms. See 1 Peter 2:9,10.

who brought them out of Egypt, who revealed his holiness to them on the mountain and who brought them into the Promised Land, who fashioned heaven and earth and all that is in them.

And in doing so, he was true to his character as the redemptive and holy God, drawing distinctions, differentiating one thing from another, setting limits to land and sea, heaven and earth, marking out the natural realm and bringing the formless chaos to the order in which he set it. The redemptive character of the creation is clear from the narrative in Genesis 1:1ff:

> In the beginning God created the heavens and the earth. The earth was without form and void, and darkness was upon the face of the deep; and the Spirit of God was moving over the face of the waters. And God said, 'Let there be light'; and there was light. And God saw that the light was good; and God separated the light from the darkness. God called the light Day, and the darkness he called Night. And there was evening and there was morning, one day. And God said, 'Let there be a firmament in the midst of the waters, and let it separate the waters from the waters.' And God made the firmament and separated the waters which were under the firmament from the waters which were above the firmament. And it was so. And God called the firmament Heaven. And there was evening and there was morning, a second day.

According to this and other passages, it is clear that God's creating work was not conceived as a haphazard collection of disparate acts but a concerted action through which not only were things created, but they were created with a definite relationship one to another. There were *orders* of creation, things terrestrial and things celestial, things of land and things of sea, there were orders of plants and of animals, of man and beast, of male and female. Moreover, all these orders bore a definite relationship with each other, thus giving the creation an internal logic and meaning, and an intrinsic order, sharply differentiated from the void and formless chaos from which it was resolved.

It will be already clear the direction in which the argument is moving. But we might note in passing that the logic of the doctrine of creation is, in fact, the foundation of all modern science.[3] Science rests upon the coherent relationship between all things in the universe, a relationship which is both intelligible and constant. Experiments conducted in one place and one field of scientific enquiry produce results which are comprehensible and useful in conjunction with results obtained in another place and in a different field. Large areas of modern life rest and depend upon this rational unity of all that is. It is interesting that in recent advances in ecological thinking, the importance of the relationship between the orders of creation, perhaps previously underestimated, is now being taken much more seriously.

It is as well, at this stage, to acknowledge the dependence of all that follows on the argument for the ordered nature of the creation outlined above. If there are no such orders, then no principles can be derived which govern their relationships, and we can be sure of no secure basis for determining whether any action is right or wrong. The argument is strictly a Judaeo-Christian one because it rests upon God's revealed action in history rather than upon any contemporary factors. It is either true for every age or false for every age. It is not open for us to argue that it might have been true in the seventh century BC, but it no longer applies today.

Most of the Old Testament legislation can be understood as maintaining, in one way or another, the divinely instituted order by which things have been created and through which they bear the right relationship to each other. A good deal of the law is given to maintaining or restoring the principal relationship with which the Old Testament is

---

3 See T. F. Torrance, *The Ground and Grammar of Theology* (Edinburgh: T. & T. Clark, 1980), for a lucid treatment of the relationship between Christian Doctrine and Experimental Science.

concerned, namely that between Israel and God himself. This relationship subsists essentially in the faithfulness of God on one side and the faithfulness of Israel on the other. The standard set for both sides is perfection, and so it is not surprising to find that the provisions for restoring Israel after breaches of faith on her side are many and precise. Also all breaches of the proper order are referenced to God. So we find that a crime or robbery against a fellow Israelite, for example, needs to be expiated not only by a restoration to the wronged man, but also by an offering to God:

> The LORD said to Moses, 'If any one sins and commits a breach of faith against the LORD by deceiving his neighbour in a matter of deposit or security, or through robbery, or if he has oppressed his neighbour or has found what was lost and lied about it, swearing falsely – in any of all the things which men do and sin therein, when one has sinned and become guilty, he shall restore what he took by robbery, or what he got by oppression, or the deposit which was committed to him, or the lost thing which he found, or anything about which he has sworn falsely; he shall restore it in full, and shall add a fifth to it, and give it to him to whom it belongs, on the day of his guilt offering. And he shall bring to the priest his guilt offering to the LORD, a ram without blemish out of the flock, valued by you at the price for a guilt offering; and the priest shall make atonement for him before the LORD, and he shall be forgiven for any of the things which one may do and thereby become guilty.'[4]

The first five books of the Old Testament contain a variety of legislation roughly grouped according to type. Chiefly they govern the

---

4 Leviticus 6:1ff.

relationship between Israel and God, giving prescriptions for worship and sacrifice, and the relationship of individual Israelites to one another and to the world in general. There are regulations as to diet, what can be eaten and what cannot, regulations as to agriculture, the raising of children, business transactions and so on. Not all these regulations are intelligible to us, but many of them are and some are generally followed today. Some can be understood on the basis of health: regulations for the diagnosis and control of leprosy, dietary restrictions, for example, the consumption of birds of prey, crustaceans, most insects except for grasshoppers and locusts presumably are forbidden partly on health and partly on cultic grounds.[5] Some cultic and sacrificial regulations signified the separation of Israel from the contamination of sin and disobedience, especially faithlessness and idolatry. Of particular relevance to this present study, we find proper relationships within the family are maintained through the prohibition of incest, and with neighbours through the prohibition of adultery.[6] Sexual intercourse with animals is a capital offence[7] as is male homosexual practice[8], and cross-dressing is prohibited[9].

These practices were outlawed in Ancient Israel because they were contrary to the order instituted by God in his creative and redemptive work. In so far as they ignored or denied the orders of creation and the proper relationship between those orders, they can only be described as intrinsically *chaotic*. Israel had been redeemed from chaos and she is thus barred from returning to it.

---

5   See Leviticus 11:9ff.
6   See Leviticus 18:6ff.
7   See Leviticus 20:15ff., Deuteronomy 27:21.
8   Leviticus 20:13: 'If a man lies with a male as with a woman, both of them have committed an abomination; they shall be put to death, their blood is upon them.'
9   Deuteronomy 21:5: 'A woman shall not wear anything that pertains to a man, nor shall a man put on a woman's garment; for whoever does these things is an abomination to the LORD your God.'

At Variance

At one level, the argument tends to dry up in the mouth. How can we develop an argument that chaos is not the same as order? It cannot be anything but chaotic for a man to have sexual intercourse with somebody else's wife, violating the exclusivity of the marital relationship. Incestuous relationships are equally chaotic. The relationship between parents and children and between siblings is an order which precludes a sexual relationship. The same is true of relations between individuals of the same sex.

Of course, people put different emphasis on the necessity for order. At an everyday level some people live rather chaotically; sometimes they pride themselves on it. Other people like to have a well-regulated life. But neither could live at all if an undergirding order was absent. The studiously muddled depend on there being a secure order as a backdrop to their chaos. Even the difference between day and night and between the days of the week and the seasons of the year are landmarks which structure human life in a necessary way. But the area where human beings are in need of the greatest order is in their relationships with one another.[10] One of the great strengths of Jewish culture is the family, the well-being of which is dependent on the proper relationships prevailing within it, and it is not surprising that two of the Ten Commandments have a direct bearing on the conduct of family life.[11] The relationship we have with our parents is not the same as with our spouses, nor again is the relationship we have with our children or with our neighbours. They are different relationships and the Judaeo-Christian tradition maintains that the differences are vital to us and to those around us.

The principle is that the relationship with spouses is sexual, but all

---

10 Of course, from a Christian point of view, our first priority is the relationship we have with God, but we do not order that ourselves.

11 The fifth and seventh: 'Honour your father and your mother, that your days may be long in the land which the LORD your God gives you. You shall not commit adultery.' See Exodus 20:12.

other relationships are not. What are the factors which make the sexual act proper within marriage and improper elsewhere? There are theological, philosophical and practical answers to this question. First, the sexual intercourse between a married couple is a defining action of their covenant to one another.[12] It makes them unique to each other and is therefore part of the realisation of marriage vows. In the Old Testament, they have to be of different sex because of the complementary nature of their union. Men and women are physically, psychologically and emotionally made for each other, and their complementary differences are important to the sacramental aspect of marriage. In Old Testament thought, the relationship between God and Israel is at least partly understood to be like a marriage. Israel is like a bride, and God is like a bridegroom. This is the force of the Song of Songs and it is also revealed more sombrely in the prophetic enactment of Hosea who was told to take a prostitute as a bride to demonstrate God's faithfulness to his Covenant despite Israel's faithlessness.[13] God's insistence that the Covenant he has made with Israel holds from his side is demonstrated in Isaiah, when he says, 'Where is the bill of divorce with which I put your mother away?'[14] There is no divorce: God is ever faithful to Israel. Notice too that the parties to the union between God and Israel are absolutely different from, and complementary to, one another. Thus, the sacramental view of marital relations requires the parties to the marriage to be one of each sex otherwise their proper differentiation and complementary nature essential to its function are lost.

Sexual relations in marriage, apart from having a sacramental function, are, in Jewish thought, inseparable from the act of procreation. They are thus part of human obedience to be fruitful and multiply.

---

12 Ordinarily, a marriage has to be consummated: intercourse has to take place to ratify the covenant made.
13 Hosea 1:2.
14 Isaiah 50:1.

Marital sex and procreation therefore belong together by way of obedience as well as simply by cause and effect. Once that order is established, all the sexual prohibitions and regulations of the Old Testament make sense. Illicit sexual activities are illogical and disordered partly because they cannot produce children, or if they do, the offspring have a chaotic relationship to those amongst whom they are generated. To the Old Testament list of prohibited sexual liaisons which could justly be ruled out on these grounds, we might add the modern practice of surrogacy, artificial insemination by donor, and the generation of children to be raised by homosexual couples. All these, including serial marriages successively terminated by divorce, pass a legacy of chaos into the succeeding generation.

Accordingly, sexual relationships that have neither sacramental coherence nor procreational function are, by Jewish and Christian standards, ruled out on the grounds of their disorder. Can there be any justification for acts which have no biological logic, no reproductive value, no sacramental significance, or which are incoherent with the orders of creation? Only, it seems, on the grounds of pleasure, which was how the Greeks and Romans justified homosexual conduct and other practices. But what happens when reproductive or sacramental purpose is abstracted from sexual activity? A fatal disjunction occurs in which pleasure is separated from purpose. Human enjoyment is found as much in the sense of proper fulfilment as in the stimulation, and there is something degenerate about the latter, when the former is absent.

In one of the best collects of the Anglican lectionary[15] we pray for grace, *to love what God commands, and to desire what he promises.* The collect prays that we should have true pleasures, lasting joys and acknowledges that the command and promise of God are the wellspring of such pleasures. Our satisfaction and joy will come when our desires are

---

15 The collect for the 4th Sunday after Easter.

directed by God's promise and our love is set upon what he commands. The collect runs:

> *Almighty God, who alone can bring order to the unruly wills and passions of sinful men: give us grace to love what you command and to desire what you promise, that in all the changes and chances of this world, our hearts may surely there be fixed, where lasting joys are to be found; through Jesus Christ, our Lord. Amen.*

In the latest campaign for the admissibility of homosexual practice in the Church, the whole direction of this collect is reversed: we are asking God and his church to command what our passions love, and we ask God to promise what we desire. Jews and, in their turn, Christians are called upon by the word and spirit of their religion to live according to the order imposed, as they believe, upon the world through God's creative and redemptive action. In this, the Old Testament avers, true pleasures are to be found.[16]

Before we leave the Old Testament, there is one remaining task which is necessary to the argument. We need to consider why some laws in the Old Testament are absorbed with scarcely any modification into Christianity, and some are seemingly ignored or seemingly repudiated. The general principle seems to be that certain laws are honoured for all time, including the Ten Commandments, some have found their fulfilment in Jesus' self offering on the Cross, and some are confined in their applicability to the specific conditions of life and

---

16 See Psalm 16:5ff: 'The LORD is my chosen portion and my cup; thou holdest my lot. The lines have fallen for me in pleasant places; yea, I have a goodly heritage. I bless the LORD who gives me counsel; in the night also my heart instructs me. I keep the LORD always before me; because he is at my right hand, I shall not be moved. Therefore my heart is glad, and my soul rejoices; my body also dwells secure. For thou dost not give me up to Sheol, or let thy godly one see the Pit. Thou dost show me the path of life; in thy presence there is fulness of joy, in thy right hand are pleasures for evermore.'

worship in Ancient Israel. We will give some consideration to Jesus' attitude towards the law in the next chapter, but here we note that the essential framework of the law was honoured and accepted by Jesus, and he declares that he has come not to abolish the law but to fulfil it.[17] This gives some clue as to how to proceed. His work was to restore Israel and therefore to bring her from the chaos of the Exile to the peace and order of her true life with God. So the first category of law, which describes that holy and ordered life under the reign of God was not only preserved but actualised in the life of Jesus and conveyed by his teaching and example, and through the power of the Holy Spirit, to his disciples. This not only includes the Ten Commandments, but also all the commands of the Old Testament relating to sexual conduct.

The second category of law, which included the bulky provisions for daily, weekly and seasonal sacrifices, Sabbath observance and Temple rites, was fulfilled in the offering Jesus made and these provisions do not now need to be repeated. The Letter to the Hebrews sets out this argument in some detail.[18]

The third category of law mainly comprises dietary regulations and health provisions. Sabbath observance also can be included here because these particular laws provided Jews (especially in later Judaism) with a visible badge of faith through which they could identify themselves as Israelites. In his letters, St Paul shows that in the New Covenant these legal badges were replaced by the badge of faith in Jesus. They were not done away with, but replaced by the duty of faithfulness, of trust, and of following Jesus Christ.

---

17 Matthew 5:17.

18 See, for example, Hebrews 7:26f: 'For it was fitting that we should have such a high priest, holy, blameless, unstained, separated from sinners, exalted above the heavens. He has no need, like those high priests, to offer sacrifices daily, first for his own sins and then for those of the people; he did this once for all when he offered up himself. Indeed, the law appoints men in their weakness as high priests, but the word of the oath, which came later than the law, appoints a Son who has been made perfect for ever.'

In connection with Israelite law, we might also note that *consent* could not be pleaded in defence against a charge of sexual misconduct. We are familiar with the modern idea that if parties consent to an act, then within broad limits, and provided they are adults, they act within the law. But no such defence was available to an Israelite charged with unlawful sexual activity. In Israel, consent to an unlawful act was meaningless. If an act was illegal, no amount of consent could make it lawful. Consenting to it simply implicated the consenting person in the crime.

Connected with this, is the fact that sex was not merely a *private* matter in Ancient Israel. It was recognised that sexual activity had public implications even if it was conducted in private. The welfare of Israel depended to a large degree on the integrity of family life, and one of the reasons that penalties for adultery were severe was that it impugned the very building blocks of Israelite national life. The whole of Israel would be contaminated by sexual misbehaviour because of the profound disruption it caused. In our modern times, the effect of private morality on public life is generally underestimated or even ignored altogether, in spite of the evident corrosive effects of immorality on society in general. Adultery and consequential divorce are experienced as contagious and can spread through institutions with remarkable speed, and the same no doubt can be true of homosexual conduct.

The theological and legal framework of Ancient Israel formed a bulwark against these degeneracies, protecting both the public and private life of Israelites. The laws prohibiting sexual immorality were so well-founded and so indispensable to Israel's continued existence as the people of God that they became equally essential in the New Testament, as Jesus effects the restoration of his people in their relationship to God and opens that restoration to the Gentiles.

3
Jesus and the Gospels

We need to recall the situation in Israel when Jesus began his ministry and the purpose of his coming. This will place his teaching in the right context and make our deductions secure. The setting, purpose and significance of Jesus' ministry has been recently set out by N. T. Wright[1] and a brief outline will be given here of some of the main features of his argument.

Israel, at the time of Christ, was under Roman occupation and, apart from a brief period under the Maccabees, had been governed by foreign rulers since the fall of Jerusalem to the Babylonian armies in 587 BC. At that time many of the leading Jews of the city were taken into exile and had to live out their faith in a foreign clime, without the visible assets of the City of Jerusalem, the Temple, the Messiah King, and far from the Promised Land, even the soil of which was considered holy. Even when they were allowed to return, Israel remained under Persian, then Greek, and finally under Roman rule. On returning to the City, of course, they could rebuild the Temple, resume their worship and sacrifices, and they had some sort of monarch, although the rule of the Herods, under Rome, was a far cry from the great days of Israelite sovereignty under King David. Nonetheless, the Jews took up their calling to live as the people of God as best they could, although there was some debate as to the best way of doing this.

---

1 N. T. Wright, *The New Testament & the People of God* (London: SPCK, 1992), and also *Jesus and the Victory of God* (London: SPCK, 1996).

Should Israelites co-operate with the Romans, or tolerate them, or try to throw them out by force? Or perhaps try to ignore or forget them amid redoubled attempts at personal holiness, as the Pharisees prescribed? It was all very difficult. The hourly reminder of foreign control, with all the pollution and indignity it brought, was provided by the squads of Roman soldiers tramping through the streets and imposing burdens upon the populace at will. Occasionally, the Roman Pro-consul would enact a decree which ordered the desecration of the Temple or other outrages, and the imposition of huge taxes and violent means of collection were a painful reminder that Israel was not at all free. In fact, in everything but geography she was still far away from her heritage, still in Exile, but still the People of the Covenant.

A great part of the Old Testament is given over to understanding why this Exile had been allowed by God, and the overwhelming conclusion was that Israel had not trusted God nor properly obeyed his commandments. There were times in the past when they tried hard to make amends, but it was of no avail: foreign control persisted, although, miraculously it seemed, their existence as the people of God was never really imperilled. In fact, numerically and economically they generally flourished and won the respect of the nations who governed them, although in the case of the Romans it was a grudging respect for their obstinacy and unwillingness to submit to or accept Roman customs. That in itself was to their credit: they refused to embrace the idolatries of their worldly masters.

With hindsight, it is possible to see Israel in something of a cleft stick. On the one hand, she was the People of God, the Covenant Nation, the Light of the World, called to holiness of life and national righteousness. On the other, she was conquered, overthrown and dominated by pagan powers, unable to rise to her calling and destiny, but kept from falling into oblivion by the hand of God upon her. It must have been a bewildering problem for the Jews who faced it.

The way in which this deadlock was broken is the story of the four

Gospels. From an Israelite point of view, it was broken in an unmistakeably Jewish, but nevertheless unimaginably radical way, in the person of Jesus Christ. Single-handed he inaugurated and effected the return of Israel from Exile. He did this, under the command and with the upholding of God the Father, by identifying himself not only with Israel, but *as* Israel though his baptism in the Jordan. He was embodying the people of Israel and doing something for them that they could not do for themselves. First, in his own wilderness wandering, he rejected all false ways, all idolatries, all vainglory, all shortcuts and all evil partnerships in the Great Return which he was accomplishing. Then he began to gather Israelites to accompany him on the return to God from Exile and all his mighty works are signs of this process.

What are the things which held Israel in Exile? The greatest of these is referred to as *hardness of heart*.[2] Hardness of heart was the attitude of the refusenik, the inclination to decide against God, to reject him on principle. This is the very issue which God promises to address in his words to Ezekiel in the exilic period. He says:

> And I will vindicate the holiness of my great name, which has been profaned among the nations, and which you have profaned among them; and the nations will know that I am the LORD, says the Lord GOD, when through you I vindicate my holiness before their eyes. For I will take you from the nations, and

---

[2] For a typical engagement of Jesus with the problem of hardness of heart, see Mark 3:1ff.: 'Again he entered the synagogue, and a man was there who had a withered hand. And they watched him, to see whether he would heal him on the Sabbath, so that they might accuse him. And he said to the man who had the withered hand, "Come here." And he said to them, "Is it lawful on the Sabbath to do good or to do harm, to save life or to kill?" But they were silent. And he looked around at them with anger, grieved at their hardness of heart, and said to the man, "Stretch out your hand." He stretched it out, and his hand was restored. The Pharisees went out, and immediately held counsel with the Hero'di-ans against him, how to destroy him.'

gather you from all the countries, and bring you into your own land. I will sprinkle clean water upon you, and you shall be clean from all your uncleannesses, and from all your idols I will cleanse you. A new heart I will give you, and a new spirit I will put within you; and I will take out of your flesh the heart of stone and give you a heart of flesh. And I will put my spirit within you, and cause you to walk in my statutes and be careful to observe my ordinances.[3]

It is significant that Jesus addresses Israel at the level of the heart, which in Jewish thought is the centre of motive and decision. The great remedy for a hard heart is repentance, the decision to accept God's judgement on things and the will to live accordingly.

Jesus' ministry consists, in very large measure, in the removal of all things which would otherwise disqualify Israelites from accompanying him on the Return from Exile. He cures infectious illnesses like leprosy, because such contamination would disqualify the sufferer from worship in the congregation of Israel. He forgives sins, much to the consternation of those who were concerned to defend the proper procedures for sin removal.[4] But of great significance is his attitude towards the principal signs of God's Covenant with Israel. He has a certain respect for the Temple, the sacrifices, the Torah and so forth. But he acts towards them in a way that can only be described as deliberate resolve to replace them with himself. 'Destroy this Temple,' he says, 'and I will raise it up in three days.'[5] 'You search the Scriptures,' he says to the Pharisees, 'because you think that in them you have eternal life; and it is they that bear witness to me.'[6] The reason why the Pharisees take exception to his forgiving of sins is that, in so doing, he short-circuits

---

3 Ezekiel 36:23ff.
4 See the healing of the paralytic, Mark 2:1ff.
5 John 2:19. See also Matthew 26:61 where witnesses accuse Jesus of saying this.
6 John 5:39.

the sacrificial system. And there is every evidence that he does so deliberately. In public, he does not refer to himself as the Messiah, for to do so would invite the mistaken idea that he was going to raise an army and try to rout the Romans. But when St Peter makes his famous affirmation at Caesarea Philippi, Jesus tells him that his declaration has been prompted by God the Father.[7] According to this, Jesus evidently accepted his own Messiahship. Finally, he substitutes himself for the whole sacrificial system in his self-offering on the Cross. In all these respects he was the fulfilment of Israel as he enacted the return from Exile for which his countrymen longed, but for whom it was impossible in their weakness.

We must now trace the impact of Jesus' ministry on Old Testament Law. We have already seen that he intended not to abolish the law but to fulfil it, and it is clear that his sacrifice on the Cross obviated the sacrificial ceremonial law and, since he replaces the Temple as the meeting place between God and Man[8] and made the final reconciliation between them, expiatory offerings no longer need to be made.

Secondly, and this is made particularly clear in the writings of St Paul, Jesus became the badge of faith for all who followed him. Christian life is life spent in the company of Jesus. After he rose from the dead, the Apostles found that he was with them in the upper room and even after his Ascension into heaven, they were assured of his presence by the power of the Holy Spirit. A transformation took place from the faith of late Judaism to the faith of the early Christians. No

---

7 Matthew 16:13.
8 'Man' is the correct term here. In the phrase 'God and Man' it means the human race from the standpoint of its relationship to God and its eternal destiny. 'Man' here means the human race in its dealings with God. In this sense the Son of God became Man, not just a human being, or just human. He became the Archetype of our race. With all proper respect to gender inclusion, 'Humankind' has inescapable resonances with biological and natural historical estimates of humanity which are not what are needed here.

longer was faith a defensive thing: a matter of warding off the pollution of the Gentiles, or of demonstrating the special calling and divine appointment of a people whose recent history seemed to argue the opposite. With the loss of their national sovereignty and with the pressure to conform to the social norms of their gentile overlords, Jews took refuge in the visible signs of their faith: notably the food regulations and Sabbath observance. The coming of Jesus marked a sea change: quite suddenly, followers of Jesus emerged boundlessly confident under the reign of their risen Lord. All the old joyful optimism of the heyday of Israelite national life was once again in evidence, except there was now no need for Land and City, Temple and Sacrifice. The new community of Jesus was complete with his presence alone.

This explains why the food laws and Sabbath observance were eclipsed in the New Covenant. Indeed, Jesus anticipates this. Confronting those who set a great store by marking themselves off from gentile contamination through dietary laws, he tells them a parable and gives a further explanation to his disciples:

> And he called the people to him again, and said to them, 'Hear me, all of you, and understand: there is nothing outside a man which by going into him can defile him; but the things which come out of a man are what defile him.' And when he had entered the house, and left the people, his disciples asked him about the parable. And he said to them, 'Then are you also without understanding? Do you not see that whatever goes into a man from outside cannot defile him, since it enters, not his heart but his stomach, and so passes on?' (Thus he declared all foods clean.) And he said, 'What comes out of a man is what defiles a man. For from within, out of the heart of man, come evil thoughts, fornication, theft, murder, adultery, coveting, wickedness, deceit, licentiousness, envy, slander, pride, foolish-

ness. All these evil things come from within, and they defile a man.'[9]

The first point to notice is that the evangelist inserts a comment on the parable being a declaration of all foods to be clean. It is striking that he understood Jesus to have the authority to make such a declaration. In other places in the Gospels, Jesus uses that authority explicitly[10] to heighten the demand of the law and apply it to the inward motive and decision of the heart, as well as the outward actions. He applies Old Testament Law more strictly on matters such as murder, adultery and divorce, while, for the reasons outlined above, relaxing the food regulations.

There is good reason for this heightening of moral standards in Jesus' ministry of returning Israel from Exile. If the law prevailed when Israel was wandering far away, how much more will it apply now she has returned to God? If that great list of wrongs: *evil thoughts, sexual immorality, theft, murder, adultery, coveting, wickedness, deceit, licentiousness, envy, slander, pride, foolishness*, were forbidden to Israel when she was in Exile, how much more inappropriate are they to the new life which Jesus brings? Jesus came to save his people from their sins.[11] It is not surprising that sins are ruled out in the Kingdom he establishes.

9 Mark 7:14ff.
10 See the Sermon on the Mount where Jesus declares, 'You have heard that it was said to the men of old, "You shall not kill; and whoever kills shall be liable to judgement." But I say to you that every one who is angry with his brother shall be liable to judgement' (Matthew 5:21), and 'You have heard that it was said, "You shall not commit adultery." But I say to you that every one who looks at a woman lustfully has already committed adultery with her in his heart. If your right eye causes you to sin, pluck it out and throw it away; it is better that you lose one of your members than that your whole body be thrown into hell. And if your right hand causes you to sin, cut it off and throw it away; it is better that you lose one of your members than that your whole body go into hell' (Matthew 5:27–30).
11 See Matthew 1:21: 'you shall call his name Jesus, for he will save his people from their sins.'

For the purposes of our present study, we need to examine some of the words used by Jesus here. The word translated as sexual immorality, *porneia*, means any and all of the forbidden sexual activities of the Old Testament. It includes incest, rape, bestiality and homosexual conduct. The word translated as licentiousness, *aselgeia*, means open and shameless immorality. The two together suggest that both secret or discreet acts of immorality and a brazenly open immoral life are equally forbidden. It is true that homosexual conduct is not mentioned specifically in this list, but the only sound way to translate the words is to afford them the meaning which would come most naturally to the speaker and the hearer in first-century Judaism. It is inconceivable that either speaker or hearer would ascribe anything but Old Testament sexual vices to these terms. Thus, homosexual conduct would most certainly be included in their meaning. The fact that *porneia* is not broken down into its constituent components reminds us that no special pleading is possible for one manifestation of *porneia* over the rest. We cannot say that homosexual conduct is admissible while incest is to be deplored. Conversely, if we want to exclude incest, then homosexual conduct must go with it.

It is sometimes thought that the Old Testament is severe and the New Testament more relaxed on the question of moral standards. On careful examination of the texts, the reverse is shown to be the case, but the thought persists because while the *demand* of the law is heightened in the New Testament, the *punishments* are remitted. It is the sacrifice of Christ which does away with the latter, it is the reign of Christ which brings about the former. The Church of God is not governed by a regime of punishment (although it does stand under the judgement of God), but it is under the rule and command of Jesus the Son of God. There is no sign whatever that Jesus relaxed, by the slightest degree, the obligations of Old Testament law with respect to sexual conduct.

But it would be wrong to think of Jesus primarily as a legislator.

The Gospel narratives give us an idea of what it was like to encounter Jesus in everyday circumstances. Such an encounter would be clearly unforgettable. It is true that Jesus confronted those to whom he came with the infinite *demand* of God. He would have their whole heart, their entire devotion. All things that obstructed this must be put away and whole heart, mind and soul devoted to following him. The rich young ruler had to sell all that he had, give to the poor and follow Jesus.[12] Would-be followers had to divest themselves of all things contrary to life in the Kingdom which Jesus inaugurated. Jesus said: 'If your right eye causes you to sin, pluck it out and throw it away; it is better that you lose one of your members than that your whole body be thrown into hell. And if your right hand causes you to sin, cut it off and throw it away; it is better that you lose one of your members than that your whole body go into hell.'[13] No one who met Jesus could fail to see the radical claim he made upon their lives.

But they could also not fail to recognise the depth of compassion, Jesus' *infinite acceptance* of them as the object of his saving work. Perhaps they had experienced the severity of governors who were not inhibited in making demands on them but who cared little for their welfare. Perhaps (though this is rather less likely) they had experienced the easy-going affection of leaders who, nevertheless, could not rise to the moral emergency in which they found themselves. Jesus was unique in that, together with the intensity of the demand, they found a love which begot an absolute commitment to their true welfare. The Apostles testify that 'grace and truth came through Jesus Christ'[14] and 'of his fulness have we all received, grace upon grace.'[15]

The third quality of Jesus Christ which distinguished him from every other person was the *sufficiency of his reigning power*. He encoun-

---

12 Matthew 19:16ff.
13 Matthew 5:29f.
14 John 1:14,17.
15 John 1:16.

tered every situation not simply as one bearing the demand of God or even the love of God, but also the regnant power of God. Nothing was impossible for him, no situation beyond his reach, no illness or disability which would not succumb to his healing word, no evil principle which would not give way to his stern command.

It is the sum of these aspects of his nature which inspired trust in those whom he met. In holding fast to principle, his compassion was not in the least diminished. And in his compassion, he had the power to amend and restore the lives of those who stood in abject need of his help. To further our present study we will examine two instances in which the grace and truth he brought into the world[16] bear on those in physical, moral and spiritual need.

We do not have an example of Jesus meeting a person guilty of *porneia* in its specific form of homosexual conduct. But we do have the story of the woman caught in adultery which prompts our consideration.

> Early in the morning he came again to the temple; all the people came to him, and he sat down and taught them. The scribes and the Pharisees brought a woman who had been caught in adultery, and placing her in the midst they said to him, 'Teacher, this woman has been caught in the act of adultery. Now in the law Moses commanded us to stone such. What do you say about her?' This they said to test him, that they might have some charge to bring against him. Jesus bent down and wrote with his finger on the ground. And as they continued to ask him, he stood up and said to them, 'Let him who is without sin among you be the first to throw a stone at her.' And once more he bent down and wrote with his finger on the ground. But when they heard it, they went away, one by one, beginning with the eldest,

---

16 John 1:15,17.

and Jesus was left alone with the woman standing before him. Jesus looked up and said to her, 'Woman, where are they? Has no one condemned you?' She said, 'No one, Lord.' And Jesus said, 'Neither do I condemn you; go, and do not sin again.'[17]

This was, of course, a trap which exploited the age-old conflict between upholding the law and showing kindness. No doubt they had heard that Jesus was overwhelmingly kind to those he met. So here was an opportunity to put his moral integrity to the test. But Jesus was not caught out. Whatever the law says, it says to everyone, and those who had gathered to execute judgement upon this hapless woman were also under the law. Jesus turns it back on them. After saying what he did, anyone who threw a stone at her, would be asserting his own sinlessness. So, one by one, they all depart. We may note in passing that when Jesus exercised judgement all other judges are displaced: the case is entirely his. We may also note that he is distracted. The woman is placed in the midst, and presumably all eyes are fixed upon her in malevolent condemnation, but not his. He thus distances himself from the malice of the crowd and is not drawn into it. They cannot recruit him.

When they have all gone away, he raises his eyes to the woman. 'Has no-one condemned you?', he asks. They could not, not because the woman was not guilty, but because in the presence of Jesus they had, even unwillingly, to cede judgement to him. His final verdict has a severe compassion about it. He does not condemn the woman, but forgives her and enjoins her sternly not to sin again. He does not excuse her. He does not offer mitigating circumstances, he does not say that adultery is not such a great sin, or perhaps not a sin at all. He does not relieve her of the obligation not to sin in the future. What he releases her from is the guilt and moral liability of her past sin.

17 John 8:2ff.

The second passage is also from John's Gospel. It is the story of the healing of the paralytic in Solomon's porticoes.

> Now there is in Jerusalem by the Sheep Gate a pool, in Hebrew called Beth-za'tha, which has five porticoes. In these lay a multitude of invalids, blind, lame, paralysed. One man was there, who had been ill for thirty-eight years. When Jesus saw him and knew that he had been lying there a long time, he said to him, 'Do you want to be healed?' The sick man answered him, 'Sir, I have no man to put me into the pool when the water is troubled, and while I am going another steps down before me.' Jesus said to him, 'Rise, take up your pallet, and walk.' And at once the man was healed, and he took up his pallet and walked.
>
> Now that day was the Sabbath. So the Jews said to the man who was cured, 'It is the Sabbath, it is not lawful for you to carry your pallet.' But he answered them, 'The man who healed me said to me, "Take up your pallet, and walk."' They asked him, 'Who is the man who said to you, "Take up your pallet, and walk"?' Now the man who had been healed did not know who it was, for Jesus had withdrawn, as there was a crowd in the place. Afterward, Jesus found him in the temple, and said to him, 'See, you are well! Sin no more, that nothing worse befall you.' The man went away and told the Jews that it was Jesus who had healed him. And this was why the Jews persecuted Jesus, because he did this on the Sabbath. But Jesus answered them, 'My Father is working still, and I am working.'[18]

Jesus asks the man what, at first sight, is a strange question. 'Do you want to be healed?' He is at a place known for its healing powers; he

---

18 John 5:2ff.

has waited there for thirty-eight years, perhaps that might signify the man's intention to recover his health. But his answer might be taken to betray the opposite. The most obvious answer is, 'Yes, of course I want to be healed.' But the man does not say that. Instead, he launches into an explanation of why he cannot be healed. What may have happened is that, at the beginning, he did intend to be healed, but that intention became eclipsed over long periods of frustration, and steadily his vision for health was replaced not only by an acceptance of but a dependence on his disability. He began to draw his identity from his predicament. 'I know him,' people would say, 'he's the chap who's been waiting thirty-eight years to be healed at that pool.'

Jesus does not wait any longer: he heals the man and sends him home with his mat. He does not persuade him to want to be healed, he cures him as an act of his own sovereignty. In so doing, Jesus cures him of a woe greater than his paralysis, the woe of being identified by his illness, a kind of idolatry in which he refers his existence and life, not to the God who created him, but to the affliction that controls him. By curing him, Jesus not only relieves him of his paralysis, but, much more importantly, re-orientates his life towards God. No longer would he be described as the man who spent thirty-eight years afflicted at the pool side, but as the man restored to health by Jesus Christ. Afterwards, Jesus finds him in the Temple and gives him a solemn injunction to sin no more that nothing worse should befall him. It reminds us that we are capable of sin whether we are afflicted with disability or not, and that, as far as Jesus is concerned, the sin is the greater woe.

What are the implications for the purposes of this study? The ministry of Jesus depicted here argues against the tendency for practitioners of homosexuality to try to define themselves by their sexuality or sexual inclinations. As far as the Judaeo-Christian tradition is concerned, this a form of idolatry. The only secure grounding for our identity as human beings lies in the fact that we have been created by God in his image and redeemed from our sins by his Son. Christians are taught to take their bearings from this point only. So strongly is this argued in the New Testament, that even the age-old distinction between Jew and Gentile collapses under its weight.[19] We cannot define ourselves as gentile Christians, or Jewish Christians, far less as 'gay' Christians. We can only define ourselves as Christians, followers of Jesus Christ.

---

19 See Ephesians 2:11ff.

4
The Letters of St Paul

The references to sexual morality in the letters of St Paul can be divided into those which simply refer to lists of immoral activities not fitting for Christians to practise, and the more theological message of the Letter to the Romans in which homosexual conduct is mentioned as part of the argument. We will consider this latter argument first.

Paul begins at Romans 1:18 to set out the prevailing situation in a fallen world into which the Gospel of Jesus Christ speaks its message of salvation. The text reads as follows:

> For the wrath of God is revealed from heaven against all ungodliness and wickedness of men who by their wickedness suppress the truth. For what can be known about God is plain to them, because God has shown it to them. Ever since the creation of the world his invisible nature, namely, his eternal power and deity, has been clearly perceived in the things that have been made. So they are without excuse; for although they knew God they did not honour him as God or give thanks to him, but they became futile in their thinking and their senseless minds were darkened. Claiming to be wise, they became fools, and exchanged the glory of the immortal God for images resembling mortal man or birds or animals or reptiles.
>
> Therefore God gave them up in the lusts of their hearts to impurity, to the dishonouring of their bodies among themselves, because they exchanged the truth about God for a lie and

worshiped and served the creature rather than the Creator, who is blessed for ever! Amen.

For this reason God gave them up to dishonourable passions. Their women exchanged natural relations for unnatural, and the men likewise gave up natural relations with women and were consumed with passion for one another, men committing shameless acts with men and receiving in their own persons the due penalty for their error.

And since they did not see fit to acknowledge God, God gave them up to a base mind and to improper conduct. They were filled with all manner of wickedness, evil, covetousness, malice. Full of envy, murder, strife, deceit, malignity, they are gossips, slanderers, haters of God, insolent, haughty, boastful, inventors of evil, disobedient to parents, foolish, faithless, heartless, ruthless. Though they know God's decree that those who do such things deserve to die, they not only do them but approve those who practice them.[1]

St Paul begins by arguing that God is just in his judgement of the world, because one way or another, primordially, everyone knew that he existed and his power was manifest in the created order. It was not that, in the first place he was not known, and that disregard for him and his laws was a simple error arising out of ignorance, but that he was known and the human race rebelled against him and defiantly refused to acknowledge his authority. Ignorance of God was not a cause of their disregard for him, but, on the contrary, their disregard for God caused them to become ignorant, futile in thought and darkened in mind. They lost their rationality.

There is in the Greek text a symmetry: because they did not see fit to give God credit, God gave them up to a discredited imagination.

---

1 Romans 1:18ff.

This is entirely consistent with Jewish thought: the fear of the Lord is the beginning of wisdom.[2] A discredited mentality lies, according to Paul, at the root of all idolatry. Indeed, the worship of the creature rather than the creator is the first result of the foolish degeneracy which follows rebellion.

Both by natural consequence and the judgement of God, this rejection of divine authority and the embrace of idols resulted in perverted sexual passions in both men and women. Richard Hays, in his book *The Moral Vision of the New Testament*[3] suggests that perverted sexual passions were a kind of sacrament (so to speak) of what he calls the anti-religion of human beings who refuse to honour God as Creator:

> When human beings engage in homosexual activity, they enact an outward and visible sign of an inward and spiritual activity: the rejection of the Creator's design. Thus Paul's choice of homosexuality as an illustration of human depravity is not merely random: it serves his rhetorical purpose by providing a vivid image of humanity's primal rejection of the sovereignty of God the Creator.[4]

Perhaps 'anti-sacrament' is a better term, since sacraments make holy or sanctify, and the conduct being described here can only profane and corrupt those who exercise it. Of course, it is true that all sins are an outward sign of an inward rebellion and so all could be said to be anti-sacraments of the Fall. But St Paul sees the connection to be more apposite in the case of homosexual conduct. He explains this with the notion of an *exchange*. The rebellion of the human race

---

2 Psalm 111:10: 'The fear of the LORD is the beginning of wisdom; a good understanding have all those who practice it. His praise endures for ever!'
3 Richard B. Hays, *The Moral Vision of the New Testament* (Edinburgh: T. & T. Clark, 1996), upon which the author acknowledges heavy dependence in this section.
4 Richard B. Hays, ibid., p. 386.

consists in an exchange of the worship of the Creator with the worship of the creature. Correspondingly, the whole of life is corrupted; the fatal exchange reappears in the forsaking of natural relationships (by which otherwise the will of God would have been done in the human race being fruitful and multiplying in the earth) and exchanging them for unnatural relations (which by their very nature are unfruitful and therefore the instruments of disobedience).

Contrast this anti-sacrament with the sacrament of marriage explained by St Paul in his Letter to the Ephesians:

> Even so husbands should love their wives as their own bodies. He who loves his wife loves himself. For no man ever hates his own flesh, but nourishes and cherishes it, as Christ does the church, because we are members of his body. 'For this reason a man shall leave his father and mother and be joined to his wife, and the two shall become one flesh.' This mystery is a profound one, and I am saying that it refers to Christ and the church; however, let each one of you love his wife as himself, and let the wife see that she respects her husband.[5]

This passage is the most explicit description of a sacrament to be found in the New Testament. The argument is the reverse of that in Romans 1. Here the old rebellion against God has been overcome in Christ. The Church is reconciled[6] to God in Christ, united once again to the Source of her life, and thus can be fruitful on earth. The coherent sign of that is marriage in which man and wife are united and fulfil the command to be fruitful and multiply.

---

5 Ephesians 5:28ff.
6 See 2 Corinthians 5:18: 'All this is from God, who through Christ reconciled us to himself and gave us the ministry of reconciliation; that is, in Christ God was reconciling the world to himself, not counting their trespasses against them, and entrusting to us the message of reconciliation.'

It is to be noted that, although St Paul sees homosexual conduct as the anti-sacrament of rebellion against God, there is no evidence that he thought it was a worse sin than any others. It takes its place in various lists of sins which have no particular order of seriousness implied. Just because St Paul uses homosexuality as a particularly appropriate illustration of his main argument, does not licence the Church to single it out for vilification. What it does amply demonstrate, however, is that the Church cannot do the reverse: she cannot exempt homosexual conduct from the charge of being a sin and a practice which inherently takes the form of collaboration with rebellion against God.

St Paul states not so much that homosexual conduct will be punished by God, but that it *is* a punishment for rebellion against him. It is a *general* consequence for a *general* rebellion. It is not individualised: he does not state that specific people are more rebellious against God than others and so find themselves in the grip of unnatural lusts. It is much more a universal cosmic consequence of the Fall of the whole race. What is important here is to recognise the direction of the argument: from the Fall to its consequences and definitely not the other way round. The logic goes: primordially the human race fell to occupy a place spiritually, morally and theologically where it should not be. As a result of this there was a 'fall-out' which, according to the Biblical writers, affected the whole ontology of the universe. There are microscopic, biological results, including viruses, there are macroscopic effects of natural disasters, but above all, there is a distortion in the human race which evidences itself in manifold sins, weaknesses and corrupt predilections. These are not uniformly distributed but manifested to varying degrees and in different combinations in all people. But, most regrettably, no one is exempt altogether, everyone bears in some way or other the afflictions of the Fall.

A recognition of this will engender a forbearance and sympathy in dealings with others, but most of all, it will guard us against pushing the logic into reverse. We will not look first at the consequences of the Fall as they are manifest in a particular individual and thence diagnose a fallen-ness in that person, making the absurd deduction that because the consequences are greater in one than in another, so is the sin. Jesus taught his disciples not to do this.[7] The relevance for the present discussion is obvious: we cannot attribute an inclination to homosexual conduct to an individualised fallenness and rebellion against God. The latter are the common lot of humanity and are manifest in individuals in diverse ways.

Returning to St Paul's text, we note that the distinction between natural and unnatural sexual activity is partly borrowed from Stoic thought, according to which homosexual conduct was considered contrary to nature and therefore deprecated. This is a Greek version of the Jewish orders of creation argument, and it is not surprising that Paul exploits both, standing as he does on the threshold between the Jewish and gentile worlds.

Any reading of Romans 1:18ff. cannot avoid the plain sense that same-sex intercourse is a mark of godlessness and a consequence of human rebellion against God, but this has not prevented several authors who want to justify homosexual conduct within the Church from attempting minimising arguments to limit St Paul's teaching to a narrow set of outlawed practices, while permitting homosexual activity in general.[8] For example, it has been claimed that the condemnation applies only to sacral prostitution, to homosexual activity in the context of idolatrous worship, but this completely ignores the context of Paul's argument which concerns the universal rebellion of man and

---

[7] John 9:2, Luke 13:1ff.

[8] See the following examples of recent literature published to promote the acceptance of homosexuality in the Church: John Boswell, *Christianity, Social Tolerance and Homosexuality* (Chicago: University of Chicago Press, 1980), John J. McNeill, *The Church and the Homosexual* (Boston: Beacon Press, 1993); James B. Nelson, *Embodiment: an approach to Sexuality & Christian Theology* (Minneapolis: Augsburg, 1978).

its consequences and the fact that homosexual practice is condemned along with a wide range of other sins. J. Boswell[9] has suggested that Paul is only condemning heterosexuals who indulge in homosexual sex on the basis that these practices have been *exchanged*. This highly strained argument breaks down completely on recollection that Paul is speaking of the whole body of humanity not to some minority subgroup within it. In any case, the words which St Paul uses refer to someone who engaged in homosexual activity, not a person of this or that orientation.

It has been further suggested that 'unnatural' simply means contrary to socially acceptable practice, but this is certainly not what it means either to Stoics or to Jews. The context shows that Paul describes as unnatural things that are contrary to the design of God, rather than simply contrary to the norms of his time.

In his correspondence with the Corinthian Church, St Paul becomes exasperated because some of the members think they have risen above the necessities of the moral law through some kind of spiritual exaltation,[10] so he declares to them:

> Do you not know that the unrighteous will not inherit the kingdom of God? Do not be deceived; neither the immoral, nor idolaters, nor adulterers, nor sexual perverts, nor thieves, nor the greedy, nor drunkards, nor revilers, nor robbers will inherit the kingdom of God. And such were some of you. But you were washed, you were sanctified, you were justified in the name of the Lord Jesus Christ and in the Spirit of our God.[11]

For our purposes, the terms of the original: *pornoi, moichoi, malachoi, arsenokoitai* are of the greatest relevance. The *pornoi* are the sexually

---

9 J. Boswell, ibid.
10 See, for example, 1 Corinthians 4:8 and 5:1f.
11 1 Corinthians 6:9ff.

immoral. This is a general term, which, as we saw in the discussion on the gospels, takes its meaning from the list of sexually immoral activities listed in Leviticus and Deuteronomy. The *moichoi* are adulterers. The term *malachoi*, literally, *soft*, is Greek slang for passive partners in homosexual activity, while *arsenokoitai*, literally *males who lie together*, signifies men who engage in homosexual practice. This latter term is a Greek word not attested in ancient literature before 1 Corinthians 6 but occasionally afterwards. It has, however, been shown to be a rendering of the language of Leviticus 18:22 and 20:13, which refer to a man 'lying with a man as with a woman', and so can be properly taken as meaning a male engaging in homosexual activity.[12]

Several further observations may be made on St Paul's declaration to the Corinthians. In the first place, his references to sexual conduct are peripheral to his purpose. His concern is that the Corinthian Church should appreciate what it meant to live under the sphere of Christ's reign into which they have been brought through baptism and the anointing of the Holy Spirit. Baptism is the sacrament by which the candidates are transformed in their allegiances from all the worldly principles, from the rule of their own appetites, and the tyranny of their sins, and brought under the rule and reign of Jesus Christ, to the righteous standard of his Kingdom, and the holiness without which no one will see the Lord.[13] In the early Church, the magnitude of the step of baptism was illustrated and brought home in a number of ways, including preparation in sackcloth, a pure white surplice to wear post baptism, and baptisteries made to look like graves so that the candidate understood himself to have died with Christ and then resurrected to a new life with him. Thence, as a baptised member of the Church, his baptism stood behind him, separating by a great chasm all that he had forsaken to belong to Christ. Paul is reminding the

---

12 See Robin Scroggs, *The New Testament & Homosexuality* (Philadelphia: Fortress Press, 1983).
13 Hebrews 12:14.

Corinthian Church on which side of this great chasm stand the sins he enumerates. To indulge in these again after baptism is unthinkable because they belong to a regime which is forever closed off to the Christian.

It is of a piece with this that St Paul, once again, does not draw any distinction between homosexual conduct and any other sin. The list betrays no particular order of importance of the elements within it. To be abusive is placed alongside robbery, drunkenness and greed. Homosexual conduct is placed alongside thieving, idolatry and adultery. It is clear that Paul is not singling any of these out for special attention: they are all equally serious in the context of the discussion in that they are all precluded from life in the Kingdom of God and of his Christ.

In 1 Timothy 1:9ff. Paul gives a further list:

> Now we know that the law is good, if any one uses it lawfully, understanding this, that the law is not laid down for the just but for the lawless and disobedient, for the ungodly and sinners, for the unholy and profane, for murderers of fathers and murderers of mothers, for manslayers, immoral persons, sodomites, kidnappers, liars, perjurers, and whatever else is contrary to sound doctrine, in accordance with the glorious gospel of the blessed God with which I have been entrusted.

Here again we find reference to *pornoi, arsenokoitai,* listed alongside murderers, kidnappers, liars and perjurers with the further inclusion of 'whatever else is contrary to sound doctrine'. Here again we find the principle that certain actions are wrong not merely because they are antisocial, unacceptable or even self-destructive, but because they are contrary to the principles of Christian teaching which recognises the ordering of things under the creative and redemptive hand of God. All these things are wrong because they are chaotic, and there is no logical way that the sexual activities listed can be singled out to be excused of this charge.

The way that the theology of Romans 1:18ff. is set out, and the way in which these lists are expressed also allow us to set to rest another strange argument occasionally voiced about St Paul. This argument suggests that St Paul himself had homosexual inclinations and his opposition to the practices was motivated by his own unresolved sexual orientation. Against this is the fact that there is a clear logic and rationality about Paul's rejection of homosexual conduct, together with all other sins, in conformity with both Jewish doctrinal antecedents and the revelation of Jesus Christ in which his thought is set. There is no sign of any special pleading, nor any vilification reserved for any group or type of sinners, no sign of anything we might call homophobia in its pejorative sense. There is the overarching concern to bring a consistency to what the Church believes doctrinally, and how it lives in practice.

But this was done by St Paul as he stood on the threshold between the Jewish and the Greek world. It was clearly an important issue for the early Church that the righteousness of life, clearly understood to be essential for the people of God, should be conveyed without dilution from the Jewish provenance of the Gospel to its new-found gentile recipients. In the letter to the new gentile converts in Antioch, sent from the Jerusalem Church, the Elders write:

> For it has seemed good to the Holy Spirit and to us to lay upon you no greater burden than these necessary things: that you abstain from what has been sacrificed to idols and from blood and from what is strangled and from unchastity [*porneia*]. If you keep yourselves from these, you will do well.[14]

14 Acts 15:28f.

There is no trace of the modern idea suggested by some[15] that the Church simply picks up the morality of the surrounding society and applies it to itself. The ancient Greek world was full of *porneia* of every description and yet any gentile convert to Christianity had to renounce it all. St Paul believed that the Church should feel no need to conform to the current predilections of the age or to its moral fashions. On the contrary, it was the world which was summoned to the faith and praxis of the Church as the Gospel was proclaimed to every creature under heaven. The Church was aware of the inevitable tension that its preaching produced. St Paul tells Timothy that persecution will be experienced by all who seek to live a godly life. The reason for this is implicit in the Church's task. She has to declare, in one way or another, the Gospel of salvation, the forgiveness of sins, redemption from the futility of disbelief and disobedience. This is all very good news, but it implies that there are sins to be forgiven, disobedience to be put away and godlessness to disown. This critical implication of the Church's good news is not lost on the world. People are frequently offended by the presumption and insolence of it, however amiable and generous and serene is the Church's address to them. It was not surprising to St Paul that, from time to time, the world became irate with the Church and sought to punish it. He considered it an inevitable affliction to be suffered by Christians who could not fulfil their duty of witness without attracting and bearing persecution.

In more recent times, John Keble makes the same point:

> The sweet and amiable and useful spirit of the Gospel will always obtain for it a certain degree of favour; but further than this, people will not go; and when the *whole* gospel, the *whole* counsel and will of God is pressed upon them earnestly and without reserve, they will presently begin to be vexed and

---

15 See, for example, R. F. Holloway, *Godless Morality* (Edinburgh: Canongate Books, 1999).

angry, and, as far as God's providence allows, will in some way or another contrive to persecute its teachers. For the whole Gospel of Jesus Christ, the whole counsel and message of God, is not only kind and gentle, but it is also a strict self-denying law. It looks to people's good, not to their satisfaction, it cares not whether they are pleased or angry, provided the great end be accomplished, of leading them, practically and in earnest, to care for their souls, and love God's truth, and amend their ways accordingly. It is absurd to expect such a message to be generally popular, as if a physician were to expect that his patients should generally like the medicines he gives them.[16]

In modern times, the Church is in great danger of forgetting this altogether. It seems that Christians will do anything to avoid the affliction, even if it means adjusting the message of the Gospel to dispel any possible offence. In reality, this means removing from the public rendering of its message any suggestion that the world might have to repent of a sin. On this basis, the effort to make homosexual conduct acceptable is driven, not by some new discovery that our understanding of Scripture has been faulty up till now, but simply by a desire not to offend, nor to be afflicted by, those who clamour for the change. It seems that the Church might be persuaded to ignore or even alter the Bible to accommodate them. As we shall see later, she has appropriated from her opponents a ready means of doing this.

---

16 John Keble *Sermon for 2nd Sunday after Trinity*, in *Sermons for the Christian Year – Sundays after Trinity I–XII* (Oxford: Society of the Holy Trinity, 1878). On 1 John 3:13.

5

Some Representative Writings from the Patristic Period

The Church Fathers[1] mention homosexual conduct only a little more frequently than do the Old and New Testaments, and wherever they do so they regard it in a thoroughly negative light. Mostly the mention is somewhat in passing in the context of a discussion of a biblical text, but all the authors are unanimous in condemning what they see as the enormity of this diabolical form of lust. For example, St Cyprian, Bishop of Carthage in the mid-third century, who has a pastoral rather than philosophical bent to his writings, says:

> Oh, if placed on that lofty watch-tower you could gaze into the secret places – if you could open the closed doors of sleeping chambers, and recall their dark recesses to the perception of sight – you would behold things done by immodest persons which no chaste eye could look upon; you would see what even to see is a crime; you would see what people embruted with the madness of vice deny that they have done, and yet hasten to do – men with frenzied lusts rushing upon men, doing things which afford no gratification even to those who do them. I am

---

1 Here is meant the principal Ante-Nicene and Post-Nicene Fathers who wrote during the first five centuries of the Church's history and whose works are collected in *The Writings of the Fathers down to AD325*, ed. Alexander Roberts & James Donaldson (Peabody: Hendrickson, 1994), and *A Select Library of the Christian Church First & Second Series*, ed. Philip Schaff & Henry Wace (Peabody: Hendrickson, 1994).

deceived if the man who is guilty of such things as these does not accuse others of them. The depraved maligns the depraved, and thinks that he himself, though conscious of the guilt, has escaped, as if consciousness were not a sufficient condemnation.[2]

There are several themes here which are common to other Patristic writers also. There is a general verdict that homosexual conduct is not only depraved, but both diabolically and incomprehensibly so – even to see such things would be a crime. In the Fourth Homily of St John Chrysostom on Romans (which is reproduced in Appendix 1 of this book), the author begins with the repetition of St Paul's words in Romans 1:26, that these practices are *vile*. Unlike many modern observers, the Church Fathers are only concerned with the homosexual act. They do not consider any possible relationship between the parties – they could not do so because they cannot get past the abject disgust they experience in contemplating homosexual union. They would be dismissed in many quarters today as being irredeemably homophobic, but, in fact, they usually reserve their opprobrium for the action itself, and scarcely stop to denounce the individuals involved. Conversely, they would find the modern world absurdly timid in dealing with it. And they would be more than astonished to find the Church actually considering giving legitimate space to something so morally reprehensible and scientifically incoherent. It also never occurs to the Patristic authors that any of their hearers would dissent from their judgement: it is self-evident to the Ancient Church that these practices are vile, degenerate and absolutely inconsistent with the teaching of Scripture and the logic of the Judaeo-Christian tradition in which they stand. It is as if they assume the force of the argument set out in the previous chapters of this book, as do their hearers, and they do not have to set it out afresh.

---

2 St Cyprian, *Letter to Donatus*, I:9.

All that they say, however, reveals their dependence upon the Biblical argument. In order to illustrate this, we will embark upon a short commentary of the major points of St John Chrysostom's Homily on Romans 1:26–27.[3]

St John Chrysostom first points out that homosexual conduct does not simply arise because of sexual exigency. It is not as if the absence of available partners of the opposite sex drove the desperate towards homosexual activity. No doubt that could happen. But, says Chrysostom, it is by a wayward decision that men forsook women and turned to their own sex. And he describes this as a *monstrous insaneness*. He is saying that such an action makes no sense: it is profoundly irrational because it rejects the good order of God in favour of a meaningless chaos. St Clement of Alexandria makes the same point in a very explicit chapter in his *Paedagogus*:

> Yet, nature has not allowed even the most sensual of beasts to sexually misuse the passage made for excrement.[4] Urine she gathers into the bladder; undigested food in the intestines; tears in the eyes; blood in the veins; wax in the ear, and mucous in the nose; so, too, there is a passage connected to the end of the intestines by means of which excrement is passed off.... The clear conclusion that we must draw, then, is that we must condemn sodomy, all fruitless sowing of seed, any unnatural methods of holding intercourse and the reversal of the sexual role in intercourse. We must rather follow the guidance of nature, which obviously disapproves of such practices from the very way she has fashioned the male organ, adapted not for receiving the seed, but for implanting it.[5]

---

3   See Appendix 1 for the English text of this Homily.
4   Literally: 'Yet in truth, it is not natural for these most lustful to put seed into the course of excrement,' trans. G. J. K. Scott.
5   St Clement of Alexandria, *Paedagogus*, trans. Simon P. Wood, Bk. II, Ch. 10, Fathers of

It is thus not only morally and spiritually disordered but biologically absurd. To adopt homosexual conduct, to misuse the body to this degree and in this manner, attracts from the Church Fathers the charge of madness. And do they not have a case? Would we not think that a man who put food in his ears might not have lost his reason? Especially if he then proceeded to insist that what he was doing was entirely right only because it gave him pleasure. To suggest to him that the body could not be nourished by the food introduced in this way would not persuade him. We find he has his own way of validating internally to himself the basis of his conduct. Is this not one definition of madness: the construction of an alternative, irrational reality in the mind which prompts actions that are incomprehensible to external observers but which make some kind of bogus internal sense to the subject and, even if those actions are entirely repugnant to others, afford him some kind of distorted pleasure?

Chrysostom goes on to say that such mad pleasures are scarcely pleasures at all. Real pleasures must be founded on good order, there is as much pleasure in the good order of an action as in the action itself. Chrysostom endorses this using somewhat Stoic language: . . . *that which is contrary to nature hath in it an irksomeness and displeasingness, so that they could not fairly allege even pleasure. For genuine pleasure is that which is according to nature.*

Now, of course, it is impossible to quantify pleasure on an objective basis. But Chrysostom is not arguing that the intensity of physical pleasure is necessarily less in homosexual engagement, but that human pleasure must have a secure relationship to that which is right and true according to Judaeo-Christian revelation. He does not ask whether it feels right or good, he estimates that any pleasure involved must be constrained by the fact that the act is chaotic and meaningless. It must, therefore, have in it some irksomeness and displeasingness.

the Church Series: *Christ the Educator* (Washington D.C.: C.U.A. Press, 1954).

St John Chrysostom then, prompted by St Paul, turns to what he describes as the *exorbitancy* of unnatural lust. The point he is making is that homosexual practice was not something into which people fell, as if by accident, nor yet a power which overtook them in their frailty, but something to which they deliberately turned in an act of defiance and mutiny against God and in their own strength and power. It was, Chrysostom says, a *work. They made a business of this sin, and not only a business, but even one zealously followed up.* Is this not what we find in our modern world? We have a whole series of charities given over to the support of the 'gay community'. We have government campaigns to prevent discrimination against homosexual people, we have 'gay' pubs, helplines, chatlines, even lines of clothing specially aimed at the 'gay' market.[6] It is worth noting, in conformity with Chrysostom's observations, that we do not have adulterers' pubs, nor helplines for the arrogant, nor government initiatives to protect the gourmandising rights of those who overeat. We might fairly ask, why is homosexual conduct made a special case? The inclination to adultery, arrogance or overeating must be strong, to account for the ubiquity of these sins, and none of them are any less deeply rooted in their perpetrators than homosexual conduct is in those who practise it. All are contrary to our true nature as human beings made in the likeness of God, all are signs of our rebellion against divine rule and purpose, and all of them lead us into trouble and ill health. Even in secular thought, all are reckoned as weaknesses and sins, except homosexual conduct. Perhaps unknowingly the world is giving homosexual conduct space because it is the anti-sacrament of unbelief. Because the world is the place of unbelief, the sphere in which man operates with a practical disregard for God, this anti-sacrament must inevitably be manifest there. There is no proof of this, and it may be simply accidental that actions reckoned as sins

---

6 It cannot be held that these things arise out of a persecution of 'gays': there is every difference between a 'gay pub' and a women's refuge.

in the Judaeo-Christian tradition are given various degrees of scope in the world. Or it may be a matter of progression: the facilities and advantages afforded to homosexual practitioners may eventually be extended to cater fairly for all vices.

St John Chrysostom then turns his attention to the consequences of homosexual practice in sowing confusion and begetting enmity between the parties. The confusion arises, he says, because the actions in question turn the world upside down. All the proper relationships were distorted or inverted, and, because homosexual conduct is a vice, anyone who indulges in it abuses his partner. They inevitably do wrong to one another. At the same time, they wrong the sex which they forsake. The devil, says Chrysostom, *sundered the sexes from one another, and made the one to become two parts in opposition to the law of God.* Thus is the race threatened, by lack of progeny through natural relations, and by the judgement of God.

But, as St Paul says, this judgement does not take the form, at least as yet, of hell and punishment, but the deed itself is the punishment for the unbelief it signifies. Chrysostom says that homosexual conduct was regarded as a right of freemen, not to be conferred on slaves, the pleasure being reserved for the privileged. But this itself was a hidden punishment for it made the real shame worse, for it made the perpetrators, who not only committed the sin, but also exalted in it, all the more pitiable to those who could see aright. This too, we find in our own time. The modern habit of 'outing', of publicly declaring homosexual practice, makes the subject all the more wretched in the eyes of the right-minded.

No matter how much homosexual practitioners try to brazen out their activities in public, however much they try to demonise those who argue that their conduct is wrong, the Church Fathers continue to insist that homosexual conduct is a sin, and indeed a sin of the worst kind. St John Chrysostom argues that it is worse than murder: *For the murderer dissevers the soul from the body, but this man ruins the soul with the body. And name what sin you will, none will you mention equal to this lawlessness.* And this is not a manifestation of some extravagant and irrational hostility against these actions, but there is a logic behind it. The murderer destroys a life by bringing it to an end, the homosexual practitioner corrupts lives and vitiates their relation to God and his good order. This position is about as far away from modern perceptions as it is possible to get, but it does not rest on an arbitrary prejudice. One of the striking things about the Patristic texts is that the authors do not simply state their positions, but they argue them at considerable length. St John Chrysostom sets out the reasoning behind his assertion that homosexual conduct is a vice because he has to. In his day the Church presented its doctrines to a world which was usually very inhospitable to them, and especially so on this issue. He is setting out the doctrines of a counter-culture: Christianity, against the prevailing pagan norms with which it was surrounded. The real difference between now and then lies in the failure of nerve of the modern Church to make its moral teaching consistent with its doctrines, and its overweening desire to gain the approval for its stance from the secular world. The Patristic authors did not seek this and were, in consequence, free to give a sound rebuttal to what they saw to be perverted practices.

St John Chrysostom expands his argument that homosexual conduct is a vice by underscoring its transmuting effect on the doers. If human beings are at least partly defined by their rationality, any departure from this, makes them something other than they ought to be, it makes them something less than human: monsters in which some physical features of humanity are present but not the essential ordered framework of their mind. Chrysostom argues that such transmuting forces should be fled, just as we might flee one who had the power to turn us into dogs. It is worth noting this because it shows us St John Chrysostom's primary concern: not to fight the world, but to guide and preserve his church members. Like all the ancient authors of the Church, Chrysostom did not think that his task was to change the world (although that happened), but to pluck people from it and save them from its vices. The Ark of ancient times was often used as model for the Church in that it was a place of safety. St John Chrysostom's sermons were intended to keep it that way.

So in his last section, Chrysostom counsels his people to live consciously in the presence of God having the fear of God before their eyes. They should not go 'window-shopping' in the world, as servants sent to do business for their master might be distracted from it by wayside entertainment, but they should fulfil their Christian calling for, as Chrysostom says, *we have been sent to despatch many affairs that are urgent and if we leave those, and stand gaping at these useless things, all our time will be wasted in vain and to no profit, and we shall suffer the extreme of punishment.*

We can aptly complete this brief review with the final injunction of St John Chrysostom to his hearers: *Since then we know these things, let us lay aside the gilded raiment, let us take up virtue and the pleasure which comes thereof. For so, both here and hereafter, shall we come to enjoy great delights, through the grace and love towards man of our Lord Jesus Christ, through Whom, and with Whom, be glory to the Father, with the Holy Spirit, for ever and ever. Amen.*

6

The Church's Plight

The purpose of the foregoing chapters has been to establish beyond reasonable doubt, that, according to its primary authorities and coherent with its essential doctrines, the Church rightly rejects homosexual activity as a valid and proper activity and justly regards it as a manifestation of chaotic behaviour from which Christians have been redeemed and to which they are forbidden to return. It has been necessary to set out this argument to answer the charge that those who oppose the admissibility of homosexual practice are simply unkind bigots, and to demonstrate the reasoned force of the traditional argument.

But it may not be sufficient, at least in the short term, to change the minds of those who have already conceded space in the Church to homosexual conduct. This is because those who concede such space have usually already conceded much else, especially in the fields of Scriptural hermeneutics and authority, and, even further back, in the whole nature of Christian belief. The Church has a much greater problem facing it than simply the challenge of homosexual conduct, and that problem lies in the fact that it has already abandoned, to a very great extent, the tools it could have otherwise used to address and defeat this challenge.

More than a hundred years ago, the Church began to listen to a very specific voice, not the voice of the Good Shepherd, but an alien voice, smooth, persuasive, intellectually distinguished, which could explain very clearly to it the nature of its faith, and which seemed to understand it better than it understood itself. This is the voice of *reductionism*, which was essentially invented by the German atheist philosopher Feuerbach, and it may be appropriate to give a brief sketch of it here. In traditional Christianity, it is held that God, who is absolutely distinct from his creation and who occupies a transcendent reality quite external to it, has revealed himself in his creating and saving acts in history, and the truth of his existence, his actions and his loving purpose have been impressed upon the minds and hearts of Christian people by the evidence of the deeds themselves, by prophetic and Apostolic witness and by the inward grace of the Holy Spirit.

In Feuerbach's scheme, this is almost perfectly reversed. The 'idea of God' is, according to Feuerbach, a projection from the minds of the believers. Not the revelation of the Infinite to the finite, but the finite generating an infinite idea. According to Harvey[1], in Feuerbach's schema, the individual self realises that it is a member of a species: 'Fascinated and entranced by the perfections of the species, the self objectifies or externalises them in the idea of a perfect being. But in contemplating and revering this alienated other, in religion, the self comes to realise that this other is its own being mystified.' Shortly stated, the individual extrapolates from his observations of the qualities of the fellow members of his species, the notion of perfection and impresses it upon a virtual deity which is generated out of his own imagination. He creates god in his own image.

The next element in Feuerbach's scheme is 'feeling'. Van A Harvey continues:

> 'Feeling' (Empfindung) is the capacity of the human ego to feel both itself and the others that act upon it. It is, [Feuerbach] writes, 'the oblique case of the ego, the ego in the accusative'. Generally, it is unfettered by the reality principle; that is, either

---

[1] Van A. Harvey, *Feuerbach and the Interpretation of Religion* (Cambridge: CUP, 1997).

by reason or by the restraints of nature. It 'breaks through all the limits of the understanding, which soars above all the boundaries of Nature.' It assumes that the deepest wishes of the heart are true, that what is wished for does, in fact, exist. And since there is no deeper human wish than that the Absolute be a being with a 'sympathetic, and tender, loving nature', the feelings rush to the judgement that there must be a deity that is personally concerned with the individual.

So, having created god in his own image, the believer is said to self-generate faith in this god, a faith which wishes this god into being the Absolute with the qualities he desires it to have.

Finally, we should add yet another feature which is especially important with regard to the present study. This is the 'felicity principle', which is the assumption that the general aim of religion is to secure the welfare or felicity of humanity in general and the self in particular. Feuerbach emphasised that religion has its roots in anxiety before death, suffering, and the longing for happiness and for recognition by another. The 'faith' so described adapts itself to the wishes and requirements of the 'faithful' to optimise their circumstances according to their own judgement. So not only does the believer create this 'god', and generate 'faith' in him, he also controls the 'will' of this god according to his own estimates of what is beneficial to him as a believer.

This latter principle forms part of the basis for the so-called 'hermeneutic of suspicion' in which religious consciousness is regarded as a false consciousness, and so instead of seeking to read and understand the text as it purports to be, an attempt is made to discern a latent and hidden motivation behind what is written. This latent motivation, or ulterior motive, as we might call it, is inevitably a less worthy motive than that suggested by a plain reading of the text. It is the method employed by the serpent in the Garden: 'You will not die. For God knows that when you eat of it your eyes will be opened, and you will be like God, knowing good and evil.'[2] The suggestion is that God did not forbid the eating of the fruit for the welfare of humanity, but for reasons of self-interest, to protect himself from their potential ascendancy. Thus the story in Genesis marks the beginning of a wide river of bad theology which was given particular impetus by the more recent concept of the hermaneutic of suspicion.

Now, Feuerbach was a self-confessed atheist and an ardent promulgator of reductionist theory. Van A. Harvey writes:

> . . . Feuerbach's suspicious hermeneutics exemplifies that feature which is at one and the same time so embarrassing to the objective scholar of religion and infuriating to the religious believer: a fervour for atheism that might itself be considered evangelical. *The masters of suspicion did not regard their demystifying work primarily as an intellectual exercise; rather, they saw it as therapy.* They thought of themselves as liberators of the human spirit, and their zeal was grounded in the assumption that belief in the gods was an illness, that it was stultifying to human beings. For Marx, religion was 'false consciousness', an expression of an estranged social existence. For Nietzsche, it was a disorder of the instincts, a reaction to suffering and the longing for another, morally better, world. For Freud, religion was a collective neurosis. For Feuerbach, religion is the 'alienation' produced when the self, in the process of differentiation from others, makes its own essential nature another objectified being. For all of these atheists, as Ricoeur has observed, their aim was not solely to destroy religion; rather, they wanted to *'clear the horizon for a more authentic word, for a new reign of Truth, not only by means of a "destructive" critique, but by the invention of an art of*

---

2 Genesis 2:5.

interpreting.' Consequently, they viewed themselves in quasi-religious terms: as prophets and evangels, as denouncers of mystification and heralds of good news.³

As concerned as he was for his ideas to gain acceptance, even Feuerbach might have been surprised at the subsequent developments in the Church following the publication of his work. Quite astonishingly, substantial tracts of the European and American Protestant Churches embraced his methods and conclusions. The anti-evangelism of Feuerbach made significant inroads into theological colleges which were anxious to gain credibility within the context of secular university life. Academically, reductionism was a kind of common currency with which the Church found itself able to deal with the secular world on an equal basis. It had a certain unjustified intellectual respectability, grounded in its atheism, which enabled the Church to avoid the criticism and derision she often had encountered previously when speaking of God and his revelation. The Church could now do intellectual business with the world in a common language. What no one seemed to notice was that in this process, the Church became atheist. Many of those outside the Church approved and the faithful did not complain – they were told that 'the assured results of modern critical scholarship' made this switch necessary. So the faithful believers became faithful atheists. Thus was effected a complete sell-out by the Church on its doctrines of revelation and on the nature of God himself. Thus to a considerable degree, the Church acquiesced to the *modus operandi* of the serpent in Genesis 1. Not only claiming to know God better than he knows himself, using a reductionist model the Church found that it could make God *be* whatever it wanted him to be.

So in various modern writings[4] we find God the Father or our Lord

---

3   See Van A. Harvey, op.cit., p. 4 (my italics).
4   See, for example, J. Spong, *Why Christianity must Change or Die* (San Francisco: Harper, 1999) and *Resurrection. Myth or Reality?* (New York: Harper, 1995). Also R. F. Holloway,

Jesus Christ transmuted to 'the idea of God' or 'the mystery we call God'. Spong argues that theism (the belief in a God who acts in the world) is now outdated by modern science and therefore Christianity must be recast without it. One of the extraordinary factors in this process was the way in which the hermeneutic of suspicion became associated with intelligence, scholarship and rationality, and conversely, the hermeneutic of recollection was consigned to stupidity, ignorance and superstition.

In the space of a hundred years, the *credo ut intellegam* of Augustine was replaced by a new motto: *dubito ut intellegam*. This new motto has now completely taken over many theological seminaries and colleges, innumerable sermons and thousands of publications, so it is not surprising that in many churches people think they cannot now believe the Nicene Creed in its plain sense. What is urgently required is to demonstrate that this lack of confidence in historic Christianity is ill-founded, and that the structure which claims to eclipse traditional Christian faith is itself unstable and highly vulnerable to criticism on the other side.

But for the purposes of our present study, the impact of reductionism has been considerable on traditional Christian morality. If we apply the hermeneutic of suspicion to Romans 1:18ff., we do not read it as was shown in a previous chapter, as if it were about God and his creation, about human disobedience and the consequent corruption of human life and action, and God's resolve to address this tragedy with his saving love and power. Rather, we read it as a convoluted commentary on the workings of St Paul's mind. We cleverly avoid being duped into thinking that the plain sense has anything to tell us. As a flash of insight, we suddenly realise that what we are reading is the contorted wrestlings of a gay man who has yet to resolve the inward conflicts of

---

having published *Godless Morality*, (Edinburgh: Canongate Books, 1999), is now speaking of 'Godless spirituality'.

his sexuality with the external taboos that confront him in the tyrannical religious world in which he lives.

In a single step, this disregards both the history, faith, spiritual insight and theological reflection of the whole of Judaism and subjugates them to an entirely speculative suggestion about Paul's sexuality and psychology for which there is no evidence. The real basis of the reductionist argument lies in the outcome desired by the commentators. The intention is to justify homosexual conduct, so that the process being studied is not even St Paul's mental conflict but the self-generated and self-fulfilling intentions of the reductionist commentators.

What we are observing is the process, described by St Paul in Romans 1:18ff., taking place, not in the primordial beginnings of mankind, but actually in the modern Church. In the particular form of reductionism, the Church has not seen fit to give credit to God and it thus is given up to a discredited imagination. It thus begins to celebrate the anti-sacrament of homosexual activity, thus conducting the outward sign of its inward refusal of God's truth and order. According to St Paul, in doing this it is receiving in itself the due penalty for its error. We could have predicted this more than a century ago if we had been sufficiently astute. Once the inevitable connection is made between failing to give God credit and the discredited mentality which follows from it, the inward part of the anti-sacrament is in place. It is only a matter of time before the hideous logic of unbelief manifests itself in the outward visible sign of perversion.

As uncomfortable as this realisation is, it does open to us the paths of repentance and return. It tells us that, however necessary penitence for homosexual conduct is, it will never be enough. Paul understands the adoption of unnatural sexual liaisons as the judgement (perhaps borne by a minority) on the unbelief of everybody. It is not sufficient for the minority to repent of the judgement. Repudiating the outward visible sign of this anti-sacrament will not be enough. The repudiation of the inward unbelief, which is its counterpart, is much more important and involves the whole Church, not just a few individuals.

In other words, we should read this inclination to legitimise homosexual activity correctly. It is not merely the Church aping the world's morality. There is a bit of this, but it is not the main impulse. We should see in it, writ large, the judgement that the Church has long since ceased to think as the Church should. What passes for theology, often is not theology but an elaborately gilded anthropology carefully decked in the language of religion, even of the Gospel. The Church likes it because it does not attract secular criticism and because it satisfies the reviewers of books and papers. It allows the Church to take its dignified place with the philosophers and worldly knowledgeable and to avoid any possible embarrassment that traditional Christian belief would otherwise cause. But secretly and inwardly, God is not given credit. He is not believed in, not trusted, not loved, not served. The Church vaunts itself in his name but is only really committed to itself, and it regards even the name of God as an alias for itself.

It is here, surely, where repentance is most urgently needed. The Church needs the ambition of St Augustine, to be Christian. There was no necessity to follow Feuerbach's reductionism, the Christian faith is as true now as it has ever been. The evidence for it is just as strong, the power just as great. And, if through penitence, the Church reverses the inward sign of unbelief, then the outward sign of sexual disorder will retreat from it and the anti-sacrament will be dissolved.

7

Casualties and Consequences

It will be apparent from the previous chapter that there be can no easy fixing of the plight into which the Church has fallen, and certainly there can be no compromise, no 'third way' by which both sides of the argument can co-exist in one communion. Over the ordination of women, it was possible to conceive of what were called 'two integrities' to allow those who approved and those who disapproved of such ordination to remain members of the one Church. The question of the admissibility of homosexual conduct in the Church is absolutely different, as the foregoing chapters of this book will have indicated. There has never been an instance in the whole history of the Judaeo-Christian tradition in which an activity, deemed unanimously to be a vice in all previous ages, has been subsequently declared to be a virtue. Furthermore, there has never been a period in which an essential doctrine of the Church has been abandoned that has not been followed by repentance and the recovery of the lost truth. This was true even when virtually the whole Church became errant in matters of doctrine. The most obvious example of this is the Arian heresy, which, for reasons which, no doubt, seemed quite reasonable to their protagonists, denied the eternity of the Son of God. For short periods this doctrine held sway, only resisted by isolated, and often persecuted saints. St Hilary of Poitiers battled to preserve orthodoxy at a time when the Church was swamped by this particular error. But, in the end, orthodoxy prevailed and Arianism (which has since reared its head many times and in different guises[1]) was, at least for the time being, put away.

Such is the power of the Gospel, that it can always raise up better hearers and more faithful disciples of Jesus Christ. The faithful should draw confidence from this. To hold a false doctrine or to advocate practices that are not lawful for Christians to adopt, only, in the last resort, threatens those who hold or advocate them. Error is only a danger to the Church in so far as she does not disassociate herself from it. And if she does not do so, then she can only cease to be the Church. Properly speaking, the Church is not a casualty of these errors, because she is constantly regenerated, purified, instructed and supplied with more saints.

In the short term, however, doctrinal and moral errors do harm the Church. The faithful are disturbed, the witness is confused, the conscience seared, and the Holy Spirit grieved. Jesus said that if a kingdom was divided against itself, it could not stand. It is strange that though this is an accepted principle when applied to political parties, which, if divided, almost inevitably lose the election, the Church is often heard telling itself that its divisions (or 'diversity') are somehow part of its strength. But from apostolic times, the unity of the Church has been understood as a sign of its union with Christ who is not divided.[2] That union must be expressed in terms of doctrine, praxis, frame of mind and goal.

One of the themes of the book of Acts is the common mind and unity of spirit of the believers. A favourite word of St Luke in this regard is *homothumadon*, being of the same mind, purpose and impulse. After the Ascension we find the disciples united in this way in prayer,[3] and in visible unity of fellowship in Solomon's Portico[4], and we find

---

1 In fact, reductionism is essentially a reappearance in part of Arianism, but a more extreme form: not only does the Son of God not have a pre-existence, he does not have a genuine existence at all!

2 See 1 Corinthians 1:13 where Paul is addressing the divisions in the Corinthian Church.

3 Acts 1:14.

4 Acts 5:12.

that same expression of unity of purpose in the letter to the new gentile converts which, as we saw earlier, urges them, *inter alia*, to abstain from *porneia*, sexual immorality.[5] So from the very beginning, the Church was defined by a public and private unanimity in doctrine, prayer and morality, which was regarded as essential to it and required of those who joined it. That is not to say that, from the beginning, the *homothumadon* of the Church was not under pressure. St Paul, in his address to the Ephesian elders,[6] warns them:

> that after my departure fierce wolves will come in among you, not sparing the flock; and from among your own selves will arise men speaking perverse things, to draw away the disciples after them. Therefore be alert, remembering that for three years I did not cease night or day to admonish every one with tears. And now I commend you to God and to the word of his grace, which is able to build you up and to give you the inheritance among all those who are sanctified.

St Paul is well aware of the forces which arraign themselves against the spiritual unanimity of the Church, and he knows that they pose a very serious threat. His response is twofold: he tells the elders to be alert, following his example and recalling to mind the teaching and admonition he gave them in the three years he spent with them, and he commends them to *God and to the word of his grace*. In both of these responses the Ephesian elders are anchored to the truth they have been taught. That will be their defence against the wolves and the speakers of perverse things. It is a striking observation from history, that though the Church has been attacked again and again in this way and so often has been led astray by false teaching and perversity, yet from one end

---

5 Acts 15:25, See comment on p. 45.
6 Acts 20:29ff.

of the earth to the other, Christians still hold fast to Apostolic teaching and those very same admonitions delivered by St Paul to the Ephesian Church.

There is therefore still a *homothumadon*, still that same mind, purpose and impulse. But now that 'same mind' is threatened by 'same sex'. And it is not a question of us all agreeing to change our minds now, even if that were possible. Because the *homothumadon* extends backwards to embrace the Christian Church in history, all the way back to the Apostles. For Christians in any age, it is even more important to be of the same mind, purpose and impulse as St Paul, than it is to be united with those of their own generation. This is a dimension of what we declare that we believe when, in the Apostles' Creed, we say, *I believe in . . . the communion of saints*. We recognise and value our fellowship with all the faithful who have gone before us and especially those who have taught us the things of Jesus Christ. If we were to accept homosexual conduct as right and valid, we would necessarily break our fellowship with all the saints and Christians of history. We would be dismissing them and all they taught, and they would have to disown us. This would have to be the case. They say that homosexual conduct is a chaotic sin from which Christ has redeemed us. If we were to say it is a legitimate act of a holy life, no agreement would be possible between the two sides. All traces of *homothumadon* would be lost.[7]

Again, what would be the impact of the acceptance of the legiti-

---

[7] We distinguish, of course, between moral and doctrinal matters which are essential to our union with the Church Fathers and other factors of time and place which are not. Our fellowship is not breached because they did not live in a democracy, nor have a vote, nor that they lived in a time of history in which there was a distinction drawn between slave and freeman or citizen and non-citizen and they did not see it as their task to reform these things. Nor because they did not permit divorce on strict grounds, and some parts of the modern Church have revived the Mosaic dispensation of divorce, is our communion with the Fathers broken.

macy of homosexual conduct upon the doctrines of the Cross? Since the event itself, the Cross of Jesus has been understood as the only and sufficient offering for the sins of the whole world. The Apostles who proclaimed the death of Jesus, and those who reflected upon it though the centuries understood it to be the greatest event in history. In it, man's moral ruin expressed over thousands of years of corruption was addressed by the holiness of God in a final settlement, the depth of which has never been fully plumbed. P. T. Forsyth, in the context of a discussion of the Fatherhood of God, says:

> We put too little into Fatherhood then if we treat it simply as boundless patient, waiting, willing love. It is more than the love which accepts either beneficence as repentance or repentance as atonement and eagerly cuts repentance short thus – 'Let's say no more about it. Pray do not mention it. Let bygones be bygones.' Forgiveness, Fatherhood for the race does not mean, with all its simplicity, just a clean page and a fresh start and a sympathetic allowance for things. God does not forgive, 'everything considered'. To understand all is not to forgive all. There was more fatherhood in the cross (where holiness met guilt) than in the prodigal's father (where love met shame). There was more fatherhood for our souls in the desertion of the cross than in that which melts our hearts in the prodigal's embrace. It is not a father's sensitive love only which we have wounded, but His holy law. Man is not a mere runaway, but a rebel; not a pitiful coward, but a bold and bitter mutineer. . . . Forgiving is not just forgetting. It is not cancelling the past. It is not mere amnesty and restoration. There is something broken in which a soul's sin shatters a world. Such is a soul's grandeur, and so great is the fall thereof; so seamless is the robe of righteousness, so ubiquitous and indefectible the moral order which makes man, man. Account must be had, somewhere and by somebody, of that

holiness of God which is the dignity of fatherhood, and the soul of manhood.[8]

Upon what possible basis could we renegotiate this? Of course, if forgiveness were merely the blinking of the divine eye, perhaps it would not matter too much what was a sin and what was not. But if the Cross does teach us that a soul's sin shatters a world, how can we say now that all sin does this except this particular manifestation of *porneia*? How can we say that the indefectible moral order of man is utterly ruined and debased by all sins except *porneia*?

It is impossible to escape the conclusion that if we were to make any sin not a sin any longer, we would inevitably do violence to the doctrine of the Cross. We would have to say that Jesus only thought he was offering a sacrifice for that sin, but he was mistaken. Equally, all Christians in the past who confessed that sin were wrong to do so, were not absolved, and wasted their spiritual energies in penitence and in striving against the temptation. So we find that, as well as doing violence to the doctrine of the communion of saints, this innovation now vitiates the doctrine of the forgiveness of sins. The forgiveness of sins is now a kind of moving quantity alongside of which Christians must run to keep up. Do we repent, confess and are forgiven, or do we run for a bit to find out whether all that will be necessary after all?

The Church has always been understood, at least in part, as a safe place. We have already seen how the doctrine of baptism made the world of the baptised a no-go area for sin. This would give the baptised double safety: they found themselves amongst the penitent, the forgiven, the restored. They were thus less vulnerable to the sins of others than they were outside. Despite the many sins of Christians, this remains substantially true of the Church. Equally, Christians them-

---

8   P. T. Forsyth, *The Church & the Sacraments* (London: Independent Press, 1906).

selves were penitents, forgiven and restored. Even though subjected to temptation, they found themselves much more able to resist it, and much freer from pressure to do wrong. This freedom originates from the reign of Christ in his Church. Where his reign is recognised and honoured, there this freedom springs up. It is necessarily attenuated in a corrupt and forgetful Church, as it is correspondingly enjoyed in a faithful and Christ-centred one.

What would be the effect of authorising homosexual conduct in the Church? In the first place, the conscience would be seared of all those who still regarded it as a sin. Their penitence would be nullified and they would be open to temptation as never before. The Church for them would cease to be a safe haven and become a moral snare. The effect would be exactly the same as on married people if adultery were to be legitimated. In either case there would be a widespread increase in the practice of *porneia*, even, as in the case of the Corinthian Church[9], exceeding the boundaries of secular morality. This would, in every degree, dissolve the Church. Any right-thinking person would have the duty to avoid it. It would become a sink of depravity, unable to proclaim the Gospel, still less able to live by it, an institution inherently at odds with its true nature and purpose. And the only result of such tension is a crisis in which the Gospel finally re-emerges in a new community of Christians who repudiate the error.

But in the meantime, much is lost, albeit temporarily. We have already mentioned the effect upon Christians who seek refuge in the Church, perhaps partly because they feel an inclination towards homosexuality. The world currently tells them that they must be true to this inclination, it is part of their very self and they must not deny it, but they must take their personal bearings from it. The Christian Gospel tells them to take those bearings from Jesus Christ and disbelieve the world's evaluation. To deny yourself, in the world's eyes, is a very

---

9 1 Corinthians 5:1.

dishonest and wretched thing, but Jesus taught Christians to do it.[10] They are free to do so because what they are and the value they have as human beings and the dear children of God, is secure and guaranteed by Jesus' love for them and his saving work on their behalf. They can deny themselves completely and forget entirely about themselves because their lives are 'hid with Christ in God'.[11] But not only can they deny themselves, they *must* do so. Thereby, Christians avail themselves of the stature and dignity that only Christ can give them, and which they can only cast away if they insist on affirming themselves. So the Church does no favour to those who wish to affirm themselves via their homosexuality in co-operating with their endeavour: on the contrary, it places them on a false track from the start.

The Church will find, too, that in addition to the many spiritual costs of the admission of homosexual practice, there is a huge moral and social cost both to its members and to the society in which it lives. Children are particularly endangered by the acceptance of homosexual practice. In the first place, male homosexual conduct in the Ancient World presupposed the involvement of children as partners and the somewhat shrill modern protestation that adult homosexual conduct has nothing whatever to do with pederasty is certainly not supported by the evidence of history. It is not the case merely that we have to accept that things are very different now. They may be. The question is whether in opening the Church door to one thing, we inadvertently admit very many other things as well. At school, children can be taught the logic of heterosexual union in the context of reproduction; no such logic can be applied to homosexual conduct. Their learning, therefore, moves from the coherent to the chaotic and their education moves backwards from the scientific order of the Judaeo-Christian thought-world to an arbitrary polymorphic world-view, which can

---

10 e.g. Matthew 16:24.
11 Colossians 3:3.

only be shaped by hedonism. There is an unmistakeable anti-educational element which corresponds to the anti-scientific nature of chaotic sexual behaviour. Even ordinary adult friendships and social relationships are skewed by the acceptance of homosexual conduct. Any friendship with a person of the same sex is much more likely to raise the question of whether there is an operative sexual element within it. Ordered living enables sexual conduct to be appropriately governed by the institution of marriage and therefore be differentiated from other types of social relationship. Any admissibility of heterosexual sex outside marriage or adultery distorts and potentially corrupts other relationships between the sexes, while the admission of homosexual conduct similarly distorts and corrupts social relations between members of the same sex. Sexuality leaks out of marriage and operates powerfully in areas in which it can only engender chaos. In this sense illicit sexual conduct is properly understood to be antisocial.

But there is a final casualty caused by the admission of homosexual conduct as legitimate in the Church. We might describe this as the anti-pastoral effect. It is commonly thought only kind to authorise homosexual conduct because, it is said, it is the only way sexual love can be expressed by those so disposed towards it, and it is in some way subhuman for anyone to be deprived of sexual activity whatever their inclinations.[12] Apart from the fact that, as this book has sought to show, the Church would have to sacrifice its whole *raison d'être* to accommodate homosexual conduct, could it not be charged with unkindness, even cruelty, if through its teaching, it deprived those inclined towards homosexual conduct of the opportunity to express sexual love in this way?

In responding to this, we first note that the Church cannot argue simply that the true numbers of individuals affected is small. It is true, that where homosexual conduct is permitted, the numbers indulging in

---

12 With the exception of paedophilia against which there is hysterical opposition.

it increase and, as in Ancient Greece at certain points in its history, can reach a majority. Correspondingly, in times of its deprecation, homosexual conduct was practised by only a very small percentage of a population, perhaps those for whom such activity represented the only possible sexual expression. But as far as the Church is concerned, the numbers are irrelevant to the principle. It has a pastoral duty to its members and the task of being the light of the world to everyone else.

One of the most well-used metaphors for the Church's ministry, drawn from that of Jesus Christ, is that of the Shepherd. Indeed the use of the word 'pastoral' is derived from this metaphor. The shepherd is responsible for the welfare of the sheep[13]. He leads them to pasture where they can feed, he wards off the wolf so they can feed in safety, and he draws them back from danger when they stray. He does not simply stroke them, or pat their heads, rather, he acts decisively for their good. He dares to believe that he knows better than they what belongs to their good and he has the courage to direct them in a way in which they themselves are not particularly inclined to go. It would not occur to him to consult them: such a process would be misleading and futile, and, above all, it might incline him to do something against his better judgement and to their inevitable disadvantage.

When St Paul and St Peter exhort the elders of the Churches to tend the flock of God,[14] it is on the basis that what is good for the flock has been well-established in Apostolic teaching and it is this of which they must be ministers. The pastoral ministry is not established by some consensus with the flock, adjusted appropriately here and there according to the preference of the sheep, but it take its bearings

---

[13] Sheep are chosen for this metaphor for good reason. They are the oldest domesticated animal in the world and have become, if anything, over-domesticated, being wholly dependent upon shepherds for food, protection and for the rearing of lambs. There are no feral sheep. Without the shepherd, they die.

[14] Acts 20:29, 1 Peter 5:2.

from Jesus and the Apostles. In this way, the good of the flock is guaranteed, their safety procured and their needs met.

The Church, in its better moments, has not shied away from this responsibility. It has, until recently at least, had the courage to tell the unmarried not to fornicate, the married not to commit adultery and everyone not to indulge in sexual immorality, including homosexual conduct. It has done this, not because it seeks to be a domineering tyrant or a killjoy, but because it wills the good of those committed to its charge, and it believes that it has learned that good from Jesus Christ, the Good Shepherd. For the Church to relax this ministry out of a misplaced sympathy for the sexually errant would be to abandon its whole pastoral duty and responsibility. The flock would inevitably become scattered and prey to the wolf.

Appendix 1

St John Chrysostom's Homily on Romans 1:26–27

For this cause God gave them up unto vile affections: for even their women did change the natural use into that which is against nature: and likewise also the men, leaving the natural use of the woman, burned in their lust one towards another.

ALL these affections then were vile, but chiefly the mad lust after males; for the soul is more the sufferer in sins, and more dishonoured, than the body in diseases. But behold how here too, as in the case of the doctrines, [Paul] deprives them of excuse, by saying of the women, that 'they changed the natural use'. For no one, he means, can say that it was by being hindered of legitimate intercourse that they came to this pass, or that it was from having no means to fulfil their desire that they were driven into this monstrous insaneness. For the changing implies possession. Which also when discoursing upon the doctrines he said, 'They changed the truth of God for a lie'. And with regard to the men again, he shows the same thing by saying, 'Leaving the natural use of the woman'. And in a like way with those, these he also puts out of all means of defending themselves by charging them not only that they had the means of gratification, and left that which they had, and went after another, but that having dishonoured that which was natural, they ran after that which was contrary to nature. But that

---

1 Reproduced from *Nicene and Post-Nicene Fathers*: First Series, ed. Philip Schaff, vol.11, *Homily IV*, pp. 355–359 (Yale 1889, reprinted Peabody, Mass: Hendrickson, 1994).

which is contrary to nature hath in it an irksomeness and displeasingness, so that they could not fairly allege even pleasure. For genuine pleasure is that which is according to nature. But when God hath left one, then all things are turned upside down. And thus not only was their doctrine Satanical, but their life too was diabolical.

Now when he was discoursing of their doctrines, he put before them the world and man's understanding, telling them that, by the judgement afforded them by God, they might through the things which are seen, have been led as by the hand to the Creator, and then by not willing to do so, they remained inexcusable. Here in the place of the world he sets the pleasure according to nature, which they would have enjoyed with more sense of security and greater gladheartedness, and so have been far removed from shameful deeds. But they would not; whence they are quite out of the pale of pardon, and have done an insult to nature itself. And a yet more disgraceful thing than these is it, when even the women seek after these intercourses, who ought to have more sense of shame than men. And here too the judgement of Paul is worthy of admiration, how having fallen upon two opposite matters he accomplishes them both with all exactness. For he wished both to speak chastely and to sting the hearer. Now both these things were not in his power to do, but one hindered the other. For if you speak chastely you shall not be able to bear hard upon the hearer. But if you are minded to touch him to the quick, you are forced to lay the naked facts before him in plain terms. But his discreet and holy soul was able to do both with exactness, and by naming nature has at once given additional force to his accusation, and also used this as a sort of veil, to keep the chasteness of his description. And next, having reproached the women first, he goes on to the men also, and says, 'And likewise also the men leaving the natural use of the woman'. Which is an evident proof of the last degree of corruptness, when both sexes are abandoned, and both he that was ordained to be the instructor of the woman, and she who was bid to become an helpmate to the man, work the deeds of enemies against one another.

And reflect too how significantly he uses his words. For he does not say that they were enamoured of, and lusted after one another, but, 'they burned in their lust one toward another'. You see that the whole of desire comes of an exorbitancy which endureth not to abide within its proper limits. For everything which transgresseth the laws by God appointed, lusteth after monstrous things and not those which be customary. For as many oftentimes having left the desire of food get to feed upon earth and small stones, and others being possessed by excessive thirst often long even for mire, thus these also ran into this ebullition of lawless love. But if you say, and whence came this intensity of lust? It was from the desertion of God: and whence is the desertion of God? from the lawlessness of them that left him; 'men with men working that which is unseemly'. Do not, he means, because you have heard that they burned, suppose that the evil was only in desire. For the greater part of it came of their luxuriousness, which also kindled into flame their lust. And this is why he did not say being swept along or being overtaken, an expression he uses elsewhere; but what? working. They made a business of the sin, and not only a business, but even one zealously followed up.

And he called it not lust, but that which is unseemly, and that properly? For they both dishonoured nature, and trampled on the laws. And see the great confusion which fell out on both side. For not only was the head turned downwards but the feet too were upwards, and they became enemies to themselves and to one another, bringing in a pernicious kind of strife, and one even more lawless than any civil war, and one rife in divisions, and of varied form. For they divided this into four new, and lawless kinds. Since this war was not twofold or threefold, but even fourfold. Consider then. It was meet, that the twain should he one, I mean the woman and the man. For 'the twain', it says, 'shall be one flesh' (Genesis 2:24.). But this the desire of intercourse effected, and united the sexes to one another. This desire the devil having taken away, and having turned the course thereof into another fashion, he thus sundered the sexes from one another, and made the one to become two parts in opposition to the law of God. For it says, 'the two shall be one flesh'; but he divided the one flesh into two: here then is one war. Again, these same two parts he provoked to war both against themselves and against one another. For even women again abused women, and not men only. And the men stood against one another, and against the female sex, as happens in a battle by night. You see a second and third war, and a fourth and fifth; there is also another, for beside what have been mentioned they also behaved lawlessly against nature itself.

For when the Devil saw that this desire it is, principally, which draws the sexes together, he was bent on cutting through the tie, so as to destroy the race, not only by their not copulating lawfully, but also by their being stirred up to war, and in sedition against one another. 'And receiving in themselves that recompense of their error which was meet.' See how he goes again to the fountain head of the evil, namely, the impiety that comes of their doctrines, and this he says is a reward of that lawlessness. For since in speaking of hell and punishment, it seemed he would not at present be credible to the ungodly and deliberate choosers of such a life, but even scorned, he shows that the punishment was in this pleasure itself. (So Plato *Theaet.* p. 176, 7.) But if they perceive it not, but are still pleased, be not amazed. For even they that are mad, and are afflicted with phrenzy (cf. Soph. *Aj.* 265–277) while doing themselves much injury and making themselves such objects of compassion, that others weep over them themselves smile and revel over what has happened. Yet we do not only for this not say that they are quit of punishment, but for this very reason are under a more grievous vengeance, in that they are unconscious of the plight they are in. For it is not the disordered but those who are sound whose votes one has to gain Yet of old the matter seemed even to be a law, and a certain law-giver among them bade the domestic slaves neither to use unguents when dry (i.e. except in bathing) nor to keep youths, giving the free this place of honour, or rather of shamefulness. Yet they, however, did not think the thing shameful, but as being a grand privilege, and one too great for slaves, the Athenian people, the wisest of people, and Solon who is so great amongst them, permitted it to the free alone. And sundry other books of the philosophers may one see full of this disease. But we do not therefore say that the thing was made lawful, but that they who received this law were pitiable, and objects for many tears.

For these are treated in the same way as women that play the whore. Or rather their plight is more miserable. For in the case of the one the intercourse, even if lawless, is yet according to nature: but this is contrary both to law and nature. For even if there were no hell, and no punishment had been threatened, this were worse than any punishment. Yet if you say 'they found pleasure in it', you tell me what adds to the vengeance. For suppose I were to see a person running naked, with his body all besmeared with mire, and yet not covering himself, but exulting in it, I should not rejoice with him, but should rather bewail that he did not even perceive that he was doing shamefully. But that I may show the atrocity in a yet clearer light, bear with me in one more example. Now if any one condemned a virgin to live in close dens, and to have intercourse with unreasoning brutes, and then she was pleased with such intercourse, would she not for this be especially a worthy object of tears, as being unable to be freed from this misery owing to her not even perceiving the misery? It is plain surely to every one.

But if that were a grievous thing, neither is this less so than that. For to be insulted by one's own kinsmen is more piteous than to be so by strangers: these I say are even worse than murderers: since to die even is better than to live under such insolency. For the murderer dissevers the soul from the body, but this man ruins the soul with the body. And name what sin you will, none will you mention equal to this lawlessness. And if they that suffer such things perceived them, they would accept ten thousand deaths so they might not suffer this evil. For there is not, there surely is not, a more grievous evil than this insolent dealing. For if when discoursing about fornication Paul said, that 'Every sin which a man doeth is without the body, but he that committeth fornication sinneth against his own body' (1 Cor. 6:18); what shall we say of this madness, which is so much worse than fornication as cannot even be expressed? For I should not only say that thou hast become a woman, but that thou hast lost thy manhood, and hast neither changed into that nature nor kept that which thou haddest, but thou hast been a traitor to both of them at once, and deserving both of men and women to be driven out and stoned, as having wronged either sex.

And that thou mayest learn what the real force of this is, if any one were to come and assure you that he would make you a dog instead of being a man, would you not flee from him as a plague? But, lo! thou hast not made thyself a dog out of a man, but an animal more disgraceful than this. For this is useful unto service, but he that hath thus given himself up is serviceable for nothing. Or again, if any one threatened to make men travail and be brought to bed, should we not be filled with indignation? But lo! now they that have run into this fury have done more grievously by themselves. For it is not the same thing to change into the nature of women, as to continue a man and yet to have become a woman; or rather neither this nor that. But if you would know the enormity of the evil from other grounds, ask on what account the law-givers punish them that make men eunuchs, and you will see that it is absolutely for no other reason than because they mutilate nature. And yet the injustice they do is nothing to this. For there have been those that were mutilated and were in many cases useful after their mutilation. But nothing can there be more worthless than a man who has pandered himself. For not the soul only, but the body also of one who hath been so treated, is disgraced, and deserves to be driven out everywhere. How many hells shall be enough for such? But if thou scoffest at hearing of hell and believest not that fire, remember Sodom. For we have seen surely we have seen, even in this present life, a semblance of hell. For since many would utterly disbelieve the things to come after the resurrection, hearing now of an unquenchable fire, God brings them to a right mind by things present. For such is the burning of Sodom, and that conflagration! And they know it well that have been at the place, and have seen with their eyes that scourge divinely sent, and the effect of the lightnings from above (Jude 7).

Consider how great is that sin, to have forced hell to appear even before its time! For whereas many thought scorn of His words, by His deeds did God show them the image thereof in a certain novel way. For that rain was unwonted, for that the intercourse was contrary to nature, and it deluged the land, since lust had done so with their souls. Wherefore also the rain was the opposite of the customary rain. Now not only did it fail to stir up the womb of the earth to the production of fruits, but made it even useless for the reception of seed. For such was also the intercourse of the men, making a body of this sort more worthless than the very land of Sodom. And what is there more detestable than a man who hath pandered himself, or what more execrable? Oh, what madness! Oh, what distraction! Whence came this lust lewdly revelling and making man's nature all that enemies could? or even worse than that, by as much as the soul is better than the body. Oh, ye that were more senseless than irrational creatures, and more shameless than dogs! for in no case does such intercourse take place with them, but nature acknowledgeth her own limits. But ye have even made our race dishonoured below things irrational, by such indignities inflicted upon and by each other. Whence then were these evils born? Of luxury; of not knowing God. For so soon as any have cast out the fear of him, all that is good straightway goes to ruin.

Now, that this may not happen, let us keep clear before our eyes the fear of God. For nothing, surely nothing, so ruins a man as to slip from this anchor, as nothing saves so much as continually looking thereto. For if by having a man before our eyes we feel more backward at doing sins, and often even through feeling abashed at servants of a better stamp we keep from doing anything amiss, consider what safety we shall enjoy by having God before our eyes! For in no case will the Devil attack us when so conditioned, in that he would be labouring without profit. But should he see us wandering abroad, and going about without a bridle, by getting a beginning in ourselves he will be able to drive us off afterwards any whither. And as it happens with thoughtless servants at market, who leave the needful services which their masters have entrusted to them, and rivet themselves at a mere haphazard to those who fall in their way, and waste out their leisure there; this also we undergo when we depart from the commandments of God. For we presently get standing on, admiring riches, and beauty of person, and the other things which we have no business with, just as those servants attend to the beggars that do jugglers' feats, and then, arriving too late, have to be grievously beaten at home. And many pass the road set before them through following others, who are behaving in the same unseemly way. But let not us so do. For we have been sent to dispatch many affairs that are urgent. And if we leave those, and stand gaping at these useless things, all our time will be wasted in vain and to no profit, and we shall suffer the extreme of punishment. For if you wish yourself to be busy, you have whereat you ought to wonder, and to gape all your days, things which are no subject for laughter, but for wondering and manifold praises. As he that admires things ridiculous, will himself often be such, and even worse than he that occasioneth the laughter. And that you may not fall into this, spring away from it forthwith. For why is it, pray, that you stand gaping and fluttering at sight of riches? What do you see so wonderful, and able to fix your eyes upon them? these gold-harnessed horses, these lackeys, partly savages, and partly eunuchs, and costly raiment, and the soul that is getting utterly soft in all this, and the haughty brow, and the bustlings, and the noise? And wherein do these things deserve wonder? What are they better than the beggars that dance and pipe in the market-place? For these too being taken with a sore famine of virtue, dance a dance more ridiculous than theirs, led and carried round at one time to costly tables, at another to the lodging of prostitute women, and at another to a swarm of flatterers and a host of hangers-on. But if they do wear gold, this is why they are the most pitiable, because the things which are nothing to them, are most the subject of their eager desire. Do not now, I pray, look at their raiment, but open their soul, and consider if it is not full of countless wounds, and clad with rags, and destitute, and defenceless! What then is the use of this madness of shows? for it were much better to be poor and living in virtue, than to be a king with wickedness; since the poor man in himself enjoys all the delights of the soul, and doff not even perceive his outward poverty for his inward riches. But the king, luxurious in those things which do not at all belong to him, is punished in those things which are his most real concern, even the soul, the thoughts, and the conscience, which are to go away with him to the other world. Since then we know these things, let us lay aside the gilded raiment, let us take up virtue and the pleasure which comes thereof. For so, both here and hereafter, shall we come to enjoy great delights, through the grace and love towards man of our Lord Jesus Christ, through Whom, and with Whom, be glory to the Father, with the Holy Spirit, for ever and ever. Amen.

Biblical Index

Old Testament

Genesis
  1:1ff. 12
  2:5 59
  2:24 77

Exodus
  20:12 16

Leviticus
  6:1ff. 14
  11:9ff. 15
  18:6ff. 15
  18:22 43
  20:13 15, 43
  20:15ff. 15

Deuteronomy
  21:5 15
  27:21 15

Psalms
  16:5ff. 19
  111:10 38

Isaiah
  50:1 17

Ezekiel
  36:23ff. 24–5

Hosea
  1:2 17

New Testament

Matthew
  1:21 28
  5:17 20
  5:21 28
  5:27ff. 28
  5:29f. 30
  16:13 26
  16:24 71
  19:16ff. 30
  26:61 25

Mark
  2:1ff. 25
  3:1ff. 24
  7:14ff. 28

Luke
  13:1ff. 41

John
  1:14, 17 30
  1:15, 17 31
  1:16 30
  2:19 25
  5:2ff. 33
  5:39 25
  8:2ff. 31–2
  9:2 41

Acts
  1:14 65
  5:12 65
  15:25 66
  15:28f. 45
  20:29 73
  20:29ff. 66

Romans
  1:16 49

1:18ff. 36–7, 41, 45, 61, 62
1:26 49
1:26f. 50, 75

1 Corinthians
   1:13 65
   4:8 42
   5:1 70
   5:1f. 42
   6:9ff. 42
   6:18 80

2 Corinthians
   5:18 39

Ephesians
   2:11ff. 35
   5:28ff. 39

Colossians
   3:3 71

1 Timothy
   1:9ff. 44

Hebrews
   7:26f. 20
   12:14 43

1 Peter
   2:9, 10 11
   5:2 73

1 John
   3:13 47

Jude
   7 81

Subject Index

adultery 15, 21, 42–3, 72, 74
   woman taken in (John 8) 31–2
Arianism 64, 65
arsenokoitai (men who engage in homosexual practice) 42–3
artifical insemination by donor 18
aselgeia (licentiousness) 29
atheism 59, 60
Augustine of Hippo, St 61, 63

baptism 43–4, 69
bestiality 15, 29
Boswell, John 41, 42

chaos 15–16, 19, 44
Christian faith and life 26–7, 43–4
Christian morality 3
   distinct from secular 2, 6, 46
   and reductionism 60–1
Christian orthodoxy 1
Church
   consequences of legitimating homosexual practice in 70–4
   and error 64–5
   pastoral duty of 72–4
   and reductionism 7–8, 57, 59–63
   to expect opposition to message 46–7
   unity of 8, 65–7
Clement of Alexandria, St 50
communion of saints 67
creation 11–13
   as foundation of modern science 13
   ordered nature of 12–13
   orders maintained by Old Testament law 13–14, 15
   redemptive nature of 12
cross, doctrine of 8, 19, 26, 68–9
cross-dressing 15
Cyprian of Carthage, St 48–9

de Vaux, Roland 10
disorder *see* chaos
divorce 18, 21, 67
doubt 61

faith 5, 58
Fall, the 40–1
feeling 57
felicity principle 58
Feuerbach, L. A. 7, 57, 59, 60, 63
forgiveness of sins 8, 25, 68–9, 70
Forsyth, P. T. 68–9
Freud, S. 59

God 2, 3, 11, 57
   compassion 5, 30–1
   creative action 19
   demands 5, 30
   faithfulness 14, 17
   holiness 11, 12, 24
   as human projection 57
   power 5, 31
   redemptive love 10, 11, 12

hardness of heart 24–5
Harvey, Van A. 57–8, 59–60

Subject Index

Hays, Richard B. 6, 38
hermeneutic of recollection 61
hermeneutic of suspicion 58–9, 61
Hilary of Poitiers, St 64
Holloway, R. F. 46, 60
homophobia 3, 45
homosexual: defined 3–4
homosexual orientation 4, 34–5, 41, 42, 49, 70–1
homosexual practice 4, 18–19
    as anti-sacrament 6, 7, 38–9, 40, 52, 62, 63
    and Church Fathers 6–7, 47–55
    and church life 62, 63, 70–4
    and church unity 8, 65–7
    as consequence of sin and unbelief 40–1, 53
    and Jesus and the Gospels 22–35
    and letters of St Paul 36–42
    nature of 19, 41, 42, 50–1, 56
    and Old Testament Judaism 10–21
    personal and social consequences of 53, 54, 70–74
    relation to other sins 40, 44, 52
homothumadon (being of the same mind, purpose and impulse) 65–7
human identity 34–5, 70–1

incest 29
Israel
    experience of God 5, 10–11, 17, 23
    history 10–11, 22–3
    law *see* Old Testament law
    redemption from chaos 15–16
    return from Exile effected by Jesus 5, 24–6
    understanding of creation 11–12

Jesus Christ
    attitude to sin 25, 32, 34
    compassion 5, 30–1
    and demand of God 5, 30
    and faith 26–7
    and forgiveness 25
    and healing 25
    Messiahship 26
    and Old Testament law 19–20, 26–8
    and power of God 5, 31
    replaces Jewish institutions 25–6
    returns Israel from Exile 5, 24–6
    self-offering on cross 19, 26
John Chrysostom, St, *Fourth Homily on Romans* 49–55, 75–83 (text)

Keble, John 46–7

McNeill, John J. 41
malachoi (partners in homosexual activity) 42–3
marriage, sexual relations within 16–18, 39
Marx, Karl 59
moichoi (adulterers) 42–3

Nelson, James B. 41
Nietzsche, F. W. 59

Old Testament law
    and family relationships 15–16
    fulfilled by Christ 19–20
    maintains orders of creation 13–14, 15
    types and categories 14–15, 20
order 16

paedophilia 72
paralysed man, healing of (John 5) 33–4
Paul, St
    and homosexual practice 36–42
    unresolved sexuality of 45, 61–2
pederasty 71
pleasure 18–19, 51
polemics 3
porneia (sexual immorality) 29, 31, 44, 45, 46, 69, 70
pornoi (the sexually immoral) 42

rape 29

rationality 8
reductionism
   and Christian morality 60–1
   and Christian theology 7–8,
     60–1, 65
   and the Church 57, 59–63
   in religion 59
repentance 24, 62
Ricoeur, P. 59

sabbath observance 27
Scripture 1–2, 56
Scroggs, Robin 43
sex education 71–2
sexual immorality 29, 31, 44, 45, 46, 69, 70
sexual relations
   and consent 21
   outside marriage 18, 72
   and pleasure 18–19, 51
   and privacy 4–5, 21
   within marriage 16–18, 39
sin
   forgiveness of 8, 25, 68–9, 70
   Jesus' attitude to 25, 32, 44
   nature of 34, 44
Spong, J. 60, 61
surrogacy 18

Ten Commandments 16, 19
theology
   renewal of 8
   reductionism in 7–8, 60–1, 65
Torrance, T. F. 13

unbelief 53, 62, 63

women, ordination of 64
Wright, N. T. 5, 22